These Fragile Outposts

These Fragile Outposts. *Cape Cod Photos*

THESE FRAGILE OUTPOSTS

— A Geological Look at Cape Cod, Marthas Vineyard, and Nantucket

By Barbara Blau Chamberlain

Published by
Parnassus Imprints
Rte 6A
Yarmouth Port, Ma. 02675

Grateful acknowledgment is made to the following for permission to use copyrighted material:

Holt, Rinehart & Winston, Inc.—*The Outermost House* by Henry Beston. Reprinted by permission of the publisher.

The Johns Hopkins Press—*The Earth Beneath the Sea* by Francis P. Shepard. Reprinted by permission of the publisher.

Joseph C. Lincoln—*Cape Cod Yesterdays* by Joseph C. Lincoln. Published by Little, Brown & Co. Reprinted by permission of the Executors of the Estate of J. Freeman Lincoln.

John Wiley & Sons, Inc.—*The New England-Acadian Shoreline* by Douglas Wilson Johnson. Copyright 1925 by John Wiley & Sons, Inc. Reprinted by permission of the publisher.

Published by arrangement with
Doubleday & Co. and The Natural
History Press.

Cover Photograph by Richard C. Kelsey

Parnassus Imprints edition
published October, 1981

Library of Congress Catalog Card Number 64–13866

Printed in the United States of America
First Edition

To Mother and Dad

I cannot but observe . . . how superior must be the pleasure which the geologist derives from scenery, above that of the man who knows nothing of the mighty agencies by which the striking features of that scenery have been produced or modified. The latter derives all his pleasure from the simple beauty or sublimity of the spot. But along with that emotion, the mind of the former is stimulated and regaled by numerous rich and delightful associations. It is carried back through immense periods of past time, during which natural causes were operating to produce the scenery before him: and he witnesses in imagination that spot, assuming peculiar and widely diverse aspects; and sees how wisely each change was adapted to bring it into its present state. It may be, too, that his mind reaches forward into futurity and perceives other changes passing over the spot, no less interesting; and the necessary consequences of the unalterable laws which God has established.

EDWARD HITCHCOCK
First State Geologist of Massachusetts, 1841

Preface

A hooked peninsula and two small islands lie grouped off the Massachusetts mainland coast, sandy intermediaries between the rocky body of New England and the open Atlantic. Written plainly along the margins of these fragile outposts and inscribed across their surfaces is the story of their one hundred million years of existence; carved into their shores is the inexorable prophecy of their future.

Here on these lands—Cape Cod, Marthas Vineyard and Nantucket—you can see the last visible remnants of an ancient northern coastal plain, a storehouse of remains of creatures and forests which lived a million centuries ago. Here you can tramp across a classic series of glacial moraines and other ice-molded features unparalleled in the East; rarely do the forms of glaciated landscape stem so obviously and directly from their glacial roots. Here, in the never-ending work of the sea, you can see in action the complex and interwoven processes of land and sea, climate and life, which may create new lands and destroy old.

This Cape and these Islands have been closely tied to American history from the first surveys of the coast in the dawn of the seventeenth century, when Champlain of France, Gosnold and John Smith of England visited the Cape-Island region and carried away with them the earliest known detailed descriptions, sketches, maps and charts of parts of this area.

Less than a score of years after Gosnold, in the harbor of Cape Cod's Provincetown, the Pilgrims first dropped anchor and drew up the Mayflower Compact before exploring the tidal marshes and upland hills of the Lower Cape. Need for fresh water and open land took them across Cape Cod Bay to Plymouth; but need for wild marsh grass as cattle fodder brought large numbers back to the Cape, and thence to the Vineyard and Nantucket.

With the ever-increasing flow of settlers and visitors to the Cape and Islands there has been continuing speculation regarding the physical aspects of these sandy lands; for they are in an endless state of flux, alive with changes measured not in geologic eons, as in the more static inlands, but within lifetimes, decades, or even overnights.

As far back as 1833, Massachusetts State Geologist Edward Hitchcock published a preliminary report on the geology of Massachusetts, believed to have been the first state geology report in the nation. In it he presented an early interpretation of the formation of the Cape and Islands. Not the earliest, however, for even during the seventeen hundreds scientists and naturalists had been writing and discoursing about the natural wonders of the region. Following hard upon Hitchcock's report came the glacial theory. With this new tool of understanding, curious naturalists and geologists ever since have been probing the histories of these lands and the processes which unendingly mold their surfaces.

In this book I have made use of this accumulated scientific knowledge, not only that which directly concerns the Cape and Islands themselves, but also that which concerns the surrounding sea floor, where many clues to their histories lie drowned. To this I have added the results of personal field and laboratory investigations. Research has led, too, through historical collections and travelers' accounts, through reminiscence, legend and tradition, wherein lies a rich leaven of history to lighten the body of scientific fact.

It is my hope that these pages will answer some of the questions that must arise in the minds of those who stand beside the Mid-Cape Highway on Shoot Flying Hill and view the miles of hummocks and lake-spattered hollows, or on Nantucket's Tom Nevers Head and watch a foaming sea suck at the foundations of the land beneath their feet; who sunbathe on shores where only a breath of earth-time ago a mile-high sheet of moving glacial ice inched its way; who farm the sandy soils or fish the shoal-floored waters; who, roaming the beaches, discover the colorful beauty of waterworn pebbles; or those who wish they could catch on canvas the ruddy face of the Vineyard's Gay Head; or who leave their foot trails next to those of the rabbits in Provincetown's dune country—of all who, awakened to any natural wonder of these sea-cradled lands, stop to ask, *What makes it so?*

B.B.C.

July, 1963

Contents

PART TWO: ONLY YESTERDAY, TODAY AND TOMORROW

List of Illustrations

Part One
THE HISTORICAL STORY

There rolls the deep where grew the tree.
O earth, what changes hast thou seen!
There where the long street roars,
hath been
The stillness of the central sea.

TENNYSON
In Memoriam A.H.H.

CHAPTER I

The First Million Centuries

East and ahead of the coast of North America, some thirty miles and more from the inner shores of Massachusetts, there stands in the open Atlantic the last fragment of an ancient and vanished land. . . . The seas broke upon these same ancient bounds long before the ice had gathered or the sun had fogged and cooled. There was once, so it would seem, a Northern coastal plain. This crumbled at its rim, time and catastrophe changed its level and its form, and the sea came inland over it through the years. Its last enduring frontier roughly corresponds to the wasted dyke of the cliff.

HENRY BESTON
The Outermost House

A Preview

Cape Cod, Marthas Vineyard and Nantucket are scenic and geological misfits. On a map they appear to be a simple extension and two outriders of the mainland. But in their landscapes and their visible geology they make up a group apart from the New England to which they belong geographically. In this geological unit, each member provides fragments—some large, some minute—of a geologic past which was shared by all three. And rooted deep within that common geologic past is the present scenery of each.

Cape Cod is the largest of the three. A man-made canal technically has turned this natural peninsula into an island. In 1602 the English explorer Bartholomew Gosnold named Cape Cod for the "great store of codfish" in its waters, but he considered the Cape to be only the hook of land at Provincetown and a part of Truro. On today's maps, however, Cape Cod embraces nearly all of Barnstable County, Massachusetts. It juts out more than thirty miles eastward from the Massachusetts mainland in a sickle-like curve—"the bare and bended arm of Massachusetts," in Thoreau's popular metaphor. It contains more or less 435 square miles of land (*see endpaper map*).

Marthas Vineyard is the largest of the Cape's island neighbors. It lies south of Cape Cod, separated from it at its nearest point by about four miles of Vineyard Sound. Like Cape Cod, it was christened by Gosnold; "Martha," it is said, in honor of his infant daughter, and "Vineyard" because of its wild grapes. The Vineyard's shape is roughly triangular, with the long, flat base of the triangle its southern shore. An extra knob of land, Gay Head, bulges out from the southwest corner of the triangle; another knob, Chappaquiddick, is similarly placed on the southeast corner, but forms an island separated from the Vineyard proper by a narrow bay. Marthas Vineyard contains some ninety-six square miles of land (*see endpaper map*).

Nantucket Island, the more remote of the two largest islands, lies about fifteen miles southeast of Marthas Vineyard and some thirty-five miles east-southeast of Woods Hole on Cape Cod. The name Nantucket is said to be from an Indian word meaning "Faraway Land." The island's more or less fifty square miles of surface lie in the sea like a thick sandy crescent. Strung off its western end are two minor islands, scarcely more than shoals: Muskeget and Tuckernuck (*see endpaper map*).

Cape Cod, Marthas Vineyard and Nantucket were produced by three massive processes.

First, obscured by vast eons, came the primordial beginnings. These can be traced back to the subterranean molding of an unseen bedrock foundation. This ancient past has left no visible record on the Cape and Islands. We must dig deep beneath their surfaces to uncover it, or probe the rocky floors of neighboring waters.

On this foundation, some one hundred million years ago, began the second process, the raising of the frameworks. These were built of sea-washed sediments and the seesaw relationships of land and restless sea through past millennia. With these frameworks began our visible record. Today it is preserved to view only on Marthas Vineyard. On Cape Cod and Nantucket it lies buried beneath glacial debris. In fact, in all of New England, Marthas Vineyard alone displays this important segment of earth history. Thus, the pages which follow reconstruct these one hundred million years on the Vineyard in some detail. For this island is the spokesman for the entire group. In its past we can read the pasts of Cape Cod and

Nantucket; and its clear history reflects the hidden history of New England, indeed, that of the entire eastern seaboard, during the past million centuries.

The final and most recent construction process draped the bulk of Cape-Island form and substance over their frameworks. This was the most abrupt and dramatic of all; it was the series of southward marches of the great Ice Age glaciers. In a matter of mere thousands of years the work of a million centuries was blanketed by countless tons of debris. In a geological overnight the Cape and Islands became much as we know them today. Again, the Vineyard dominates in the early part of this process; for here we find the clearest record of the earlier ice invasions.

The present book falls into two parts. In PART ONE: THE HISTORICAL STORY we follow a natural chronological order to trace the creation of the Cape and Islands. Thus, the earliest chapters deal mainly with the Islands, especially the Vineyard. Nantucket attains status as a distinct feature toward the end of the Ice Ages. Cape Cod, for the most part, waits until the final burst of glacial activity to find its place on the map.

In PART TWO: ONLY YESTERDAY, TODAY AND TOMORROW we take a close look at the unresting decorative agencies active today on Cape-Island surfaces and shores: beach and dune; storm and current; moor and marsh; lake and forest. And finally, past and present are fused together into a window through which we view a seemingly inevitable future.

The Foundations

New England's rocky foundation, its solid bedrock crust, lies very deep below the surface in only a few places, and these include Cape Cod and the Islands. Elsewhere, in many parts of New England, bedrock is at the surface.

In this respect, most of New England—a land of ancient mountains and granite-rooted uplands—bears little resemblance to these soft earth-mounds off the Massachusetts coast. In fact, more than five hundred miles of rocky shores armor New England; only the small segment starting at Kingston, just north of Plymouth, and continuing through where the coast juts out into the Bay as Cape Cod and the Islands, is soft, sandy and vulnerable. The *Mayflower* happened to avoid entirely the "stern and rock-bound coast."

For as we near the Cape and Islands the depth to bedrock increases, and keeps increasing as we travel east across them. A few miles west of the Cape at Onset, bedrock lies sixty to one hundred feet beneath the loose covering sediments. From Onset the bedrock foundation slopes gradually seaward, so that beneath the Cape and Islands it is several times as deep. In fact, beneath most of the Cape and Islands, bedrock has remained hidden beyond the greatest depth reached by drills probing for water. These have bored through the entire thickness of sand, clay and gravel which rise in some places more than one hundred and fifty feet above sea level, and down below sea level through a greater distance of similar loose sediments. All they have had to chew upon have been clay, sand and boulders.

So deeply buried is the bedrock that when the Army Engineers started construction on the canal-spanning Buzzards Bay Railroad Bridge, tests of the subsurface materials on either shore revealed that they were too weak even to bear the necessary load of steel, and more than a hundred long oak piles had to be driven under the bases of the bridge abutments to support them.

South of Onset, on the Elizabeth Islands, which lie strung out like beads on a chain southwestward from Cape Cod's Woods Hole, wells have gone 230 feet deep and have not reached bedrock. Seismograph findings of a team of geophysicists under Maurice Ewing indicate that at Woods Hole, bedrock lies about 294 feet below sea level.

Its seaward slope carries the bedrock deeper as we go eastward along the Cape peninsula. Until the winter of 1961–1962, in fact, solid rock had been struck at one place only, in Provincetown, half a century ago. There was no proof that it really was bedrock and not merely a large boulder. But recent drilling of two thousand-foot holes on Cape Cod, in Brewster and Harwich, finally has fathomed the Cape's elusive foundation. In each case bedrock was struck at about 435 feet beneath the surface.

South of Cape Cod, geophysical findings indicate that bedrock just off Gay Head, the southwest tip of Marthas Vineyard, lies 770 feet beneath sea level. It may be several hundred feet deeper beneath Nantucket.

Yet the shell of the earth, deep though it lies beneath loose surface sediments on the Cape and Islands, is not quiet. On occasion it has announced its presence by responding to the unimaginable forces within the earth's unstable crust. It has moved and sent massive vibrations upward through the loose debris.

Thus, earthquakes, although far from common in this region, have been known to occur. The Cape and Islands are parts of a minor earthquake belt swinging up the eastern coast from the Carolinas to Canada's Maritime Provinces. Sandwich had scarcely established itself as a town in 1637 when it was shaken by an earthquake the following year. For four minutes there were claps like thunder, the ground shook, and chimneys fell. The pious inhabitants wondered if their settlement met with divine disapproval while smaller quakes continued for three weeks. The bedrock stirred and trembled slightly several more times during the nineteenth and twentieth centuries. Provincetown had a tremor in April, 1935, which rattled windows from the Cape tip southward to Wellfleet. An inhabitant of Provincetown described it as "a loud report, followed by a couple of reverberations, such as might have been made by a big gun fired at sea, or a not too distant single peal of thunder." Seismographs reported the focus, or place of origin, in the bedrock beneath the southern end of submerged Stellwagon Bank north of Provincetown.

This slightly unstable bedrock plain supports the drowned edge of the New England part of the continent. Together they form the continental shelf. The shelf ends abruptly at the continental slope, where the sea floor drops more than a mile into the abyss. All of the continents are bordered by such a continental shelf, over which seas, higher now than during parts of the geologic past, have encroached upon the land. The rock beneath the shelf surface, above which Cape Cod and the Islands rise hundreds of feet, is primarily a continuation of the tough and ancient granitic bedrock foundations of the continents. Such rock underlies the softer sediments of nearly all of the earth's land areas. In some places it shoulders its way above the surface so that mountains, such as New Hampshire's Presidential Range, dominate the landscape.

The rock-floored continental shelf and slope off New England are the dumping grounds of that part of the continent. They receive most of the millions of tons of sediment carried to the sea each year by New England's network of rivers. Such stream-carried debris —clay and mud, silt, sand and pebbles—mingles with debris snatched from the land and blown to the water by wind, and with debris eaten from New England's shores by the sea itself. With the deliberate slowness typical of the earth's doings, some of these materials accumulate near shore. Much of the finer silt, since it tends to remain suspended, drifts to the deep waters which start at the New England part of the continental slope.

In the Cape region, the continental shelf is about two hundred miles wide. In the course of these two hundred miles, the surface of its sediment blanket descends from where it enters the sea at the mainland's shore to a depth of about six hundred feet, where it reaches the slope.

The Cape-Island bedrock is a continuation of that which floors the nearby bodies of water. Beneath Massachusetts Bay, Cape Cod Bay, Buzzards Bay and the Sounds, there are several rock types. Some date back to a primeval, lifeless earth. Others came into being only a short time before the embryo Cape and Islands began their development, only yesterday in earth history.

Part of the underwater rock is igneous, that is, it originated as flows of hot lava or as intrusions of molten magma. Long since cooled and crystallized, this rock is light in color and relatively light in weight, and therefore granite-like, quite similar to rocks around Salem and at Rhode Island, New Bedford, and Dedham. Much of the floor of Massachusetts Bay, in fact, may well be a continuation of the granite-like Dedham rock. In contrast, the submerged bedrock in other places is dark and heavy, resembling rock which can be seen near Attleboro and Boston.

Still other parts of the offshore rock sea floor originated as layered beds of gravel, sand and clay, formed during an ancient cycle of sedimentation and dropped as loose debris east of the present mainland coast. These sediments eventually became cemented together into solid rock.

According to geophysical speculation, about a quarter of a billion years ago a relentless pressure began to act on the more ancient of these rocks. It was not a sudden, cataclysmic pressure—if it had been, the rocks would have cracked. Instead, they yielded slowly. By unwitnessed magic of crystal transformation, the subtle transmutation of one mineral into another better able to stand the strain, the rock itself changed. At the same time it folded, crumpled, lifted itself and became a chain of mountains. These mountains may have risen more than two miles high across the present location of Cape Cod and the Islands.

Then long ages of erosion took their toll. Eventually the mountains disappeared. Where they had stood, there remained only a broad plain of rock, crumpled and hardened by the pressure it had undergone. Over this plain eventually the rising ocean spread. This rock is the foundation that lies immediately beneath the loose sedimentary structures which make up Cape Cod, Marthas Vineyard and Nantucket. The two drillings recently completed on Cape Cod found bedrock of pressure-formed schist and gneiss—the anciently eroded core of the still more ancient mountains.

From this remote bedrock stage we can pick up, at the nearer milestone of one hundred million years ago, the thread of the second of the processes which formed these lands. This was when their frameworks began to rise on the rocky foundations. It is the start of the visible record of the Cape-Island ancient past. Not so ancient, however, when viewed against the estimated 4.7-billion-year history of the earth itself. Relative to the life of the earth, their entire million-century existence is scarcely more than the unnoticed graying of a hair.

The Frameworks: Gay Head's Pages of the Past

On a windy summer day in the late eighteenth century, the rough waters of Vineyard Sound were causing considerable distress to a passenger on a small ship:

> A calm . . . coming on with a hot sun and a constant rolling of the boat, I grew exceedingly sick. Nothing could alleviate my feelings but a view of Gay Head . . . at the distance of about fifteen miles. A variety of colors, such as red, yellow, and white, differently shaded and combined, exhibited a scene sufficient to captivate the mind, however distressed.

Thus arrived William Baylies, one of the first geologists to feel the impact of these mighty, mile-long sea cliffs (*Figure 1*). Sediment layers, streaked like bright finger-smeared dyes, show red, green, tan, gray, yellow, brilliant white and nearly coal black across the face of the cliffs and give this foreland its name.

Some layers are primarily sand, others mainly gravel, others compact clay. Some are full of fossils; large glacial boulders protrude like bony knuckles from others. This colorful confusion of sediments, the creased and shuffled pages of an ancient manuscript, contains New England's most complete record of its past hundred million years. It tells a story of a land first covered with ancient forests,

Figure 1. *A record of one million centuries—part of the Gay Head Cliffs, Marthas Vineyard.* U. S. GEOLOGICAL SURVEY

then engulfed by seawaters, laid bare and engulfed again; of giant sharks swimming over what are now the meadows of Chilmark, of camels and horses roaming the plains where Gay Head now rises; of a series of overpowering invasions by ice floes from the north.

Today the colored layers run in many directions, horizontal, diagonal and vertical. Fallen debris from the face of the cliffs often obscures the lower parts. Sometimes entire wedges, like huge slices of layer cake, are cut off by waves and fall to land at odd angles down the cliff, further complicating the pattern. The irregular arrangement and varied components of the debris layers partly account for the picturesque face of the cliffs today. Firm clay forms bulging headlands, projecting shelves and sharp pinnacles. Loose, softer sand is easily eroded, leaving caverns and hollows. The sea makes its way into the land more slowly where layers which present their edges to the sea are nearly horizontal than where they are vertical; for where soft crumbly materials stand vertically, the waves probe into the weak zones, making severe inroads into the cliff. This, primarily, is what has formed the amphitheater known as the Devils Den, deepened further by the commercial removal of clay during past centuries.

However confused the cliff layers may seem today, each has a definite meaning. Every layer, such as the rusty red sand or the white clay, represents a well-defined period of accumulation which may have lasted thousands, in some cases perhaps millions, of years. During each deposition period, conditions in the region remained almost the same, so that the same sort of materials continued to pile up, dropped by rivers, sea or ice. When local conditions changed, so did the materials deposited.

Naturally there also were stretches of time, within the one-hundred-million-year span here portrayed, which left nothing to show for themselves in this region. Erosion was dominant and materials were stripped away.

Originally, of course, all layers were flat, with the earliest, oldest beds at the bottom. Successively younger layers formed upon them as successively more recent river, sea or glacial activity came into action. Later, the inexorable force of huge moving ice sheets played havoc with them, pushing them up into wrinkles as easily as sheets in a stack of paper can be wrinkled by a moving hand.

Chalk-white layered clay is one of the most striking materials at Gay Head (*Figure 2*). The uppermost of the several clay layers

Figure 2. *A close-up view of Gay Head's strata, Marthas Vineyard.*
U. S. GEOLOGICAL SURVEY

is about four feet thick and is white true clay, plastic when wet and hard when dry. During the last century a small clamshell showed up in the clay, and opened a previously closed short chapter in earth history. This indicated that great changes and huge forces were parts of the story. For the small remnant of life, the clamshell, told that the clay in the bluffs today, now reached only by the ocean's salt spray, once was sea floor. Under some vanished force—presumably glacial—that sea floor crumpled, and heaved itself free of the waters to become a twisted girder in the framework of Marthas Vineyard.

And we can read back still further into time by looking at the layer below. Here we find no clamshells. Instead, there are filmy carbonaceous residues of plant leaves and other bits of ancient vegetation. Sometimes twigs, seeds and nuts lie embedded in the clay. Here and there one can see rusty, angular, clumped masses of clay with groups of fossil leaves cemented into them and with the sparkle of tiny mica flakes on their surfaces. Near the very base of the cliff, only rarely visible from the beach, are more concentrated plant remains in the form of brown and black patches of lignite, a low-grade, plant-derived fuel midway between peat and soft coal. This lignite is the earliest visible deposit of all.[1]

Ancient Forests on Marthas Vineyard

The lignite is indisputable evidence that green forests existed here. They rooted themselves in an ancient forest floor which today forms only a thin layer of the massive earth foundation supporting Gay Head's lighthouse. Since the forest remains lie beneath the clay of the ancient sea floor, we know that this region had been land before it became a sea floor, to be raised again as today's land. Now, in this small fragment of the endless worldwide cycle

[1] The dramatic ruggedness and vivid colors of the cliffs, and the strange black lignite within them, have given wings to the imaginations of Indians and inhabitants and even early geologists. In 1793, William Baylies reported on the geology of Gay Head, ascribing the whole region to the action of a volcano. A Vineyard historian reported in 1815: "Several persons who have visited Marthas Vineyard suppose that this substance [the lignite] indicates the vicinity of a coal mine; whilst others imagine that it is nothing but pieces of charcoal, made either by a volcano, or by Maushop, when he was cooking his whales." According to this account, "volcanick flames [were said to] have been seen to ascend from the Devils Den." Maushop was the legendary benign Indian giant said to dwell in the Devils Den and to broil whales on fires made of large trees.

of sea and land, we can watch the sea nibbling at the cliffs and reclaiming that land once again.

Some of the lignite is eighty percent vegetable matter. It is a one-hundred-million-year-old plant tomb in which pieces of ancient tree trunks, still recognizable, lie with reddish-yellow drops of an amber-like material which may be gum from the nuts of eucalyptus trees; bits of this gummy substance sometimes bob about in the waves which lap Gay Head's beach. There are also remnants of wood riddled with holes put there long ago by the same sort of little shipworm as that which still honeycombs pilings and ship hulls.

Careful microscopic studies of the trunks, pollens and other vegetable matter remaining to us in the lignite and associated clays have revealed that more than 117 species of plants once grew here. When they died, they were buried before their remains totally disintegrated. Could we jump back in time and stroll through that forest of the ancient Vineyard, we should see many plants resembling those that we know well today—oaks, pines and willows; holly and sassafras; magnolias and sequoias.

Unfortunately the intriguing lignite now is hidden deep at the base of the bank behind the debris of landslides, and its public appearances are brief and erratic. In the past, when it was more clearly exposed, the clear signs of life in the lignite and the associated clay enabled geologists to date these earliest known Vineyard deposits, long before most of the island's geology had been untangled. As far back as 1823, when systematic geology was still in its infancy, John Finch studied the lignite and observed:

> When we . . . arrive at those rocks which contain fossils, we find each stratum decidedly marked by the remains of . . . shells peculiar to it. These fossils constitute the medals of the ancient world, by which to ascertain the various periods, during which the exterior coat of the earth was consolidated. It has been observed, that these organized remains occur in so regular an order, that it is like examining a cabinet of shells, where you are sure to find, in every drawer, those peculiar ones which must have been deposited within it.

The fossils in these Vineyard deposits indicate a mid-Cretaceous age. This means that these beds have lain on the site where we see them some one hundred million years.

Yet even while the plants lived, the earth already was ancient. Life had been stirring on its surface for nearly four hundred million years. During the mid-Cretaceous, changes were taking place, leisurely but of tremendous moment. In the forests and on the plains roamed newcomers, small timid mammals, sharing the scene with the last of the dinosaurs. A new class of vegetation, the flowering plants, gave these mammals the food on which they depended, and with it, the gift of survival. Competing with ferns, ginkgoes and primitive evergreens for sun and space, these plants were carpeting the vast plains with grasses and gracing the forests with modern-type hardwood trees.

At that time the edge of the sea lay somewhere to the east of today's Vineyard. The Vineyard-Cape Cod-Nantucket area existed simply as a part of the foggy beginnings, now largely drowned, of New England's coastal shallows.

Through the long thousands of centuries, the Vineyard forests became swampy. Low-lying and poorly drained, the dead plants did not readily decay, nor were they removed by running water. Instead, vegetable matter piled up in a spongy, dark, pungent-smelling mass of peat.

For the sea was rising. It was one of the rises that would carry it flooding over large parts of the continent during the Cretaceous. Never since that period has so much of the earth's surface lain covered by seawater. Salt water began to encroach upon southeastern New England's swampy forests. Gradually their muddy waters became brackish.

Origin of the Cretaceous Clay

Probably the Vineyard area at this time was delta-like; rivers, flowing from highlands to the west, reached the edge of the sea here. When we look at the compact clay beds, we can picture the murky streams which wandered sluggishly into these swamps, laden with clouds of suspended clay from rock weathering on higher lands. The clay settled from the streams and spread gradually over the swamp grounds and about the bases of the plants. It built up, ever so slowly, against their stems and trunks as a thick, airtight mass. In some cases plants perished, smothered by it. Occasionally leaves and stems fell to the mud to become preserved as imprints in the clay. Sometimes during floods, sand and quartz pebbles were caught by obstacles in the stream bed and piled up, so that we find pods of coarser material in the clay today.

The chain of mountains in which these rivers started was a predecessor of today's Appalachians, extending down northeastern North America. It had risen during an earth upheaval some one hundred and twenty million years earlier, and ever since had been subject to the slow, inevitably destructive forces of weather and erosion. During these 120,000,000 years, air and water had been attacking exposed rock surfaces relentlessly. Tiny cracks in rock, wedged apart by freezing water, grew until great chunks broke off. Oxygen, water and carbon dioxide reacted chemically with hard minerals, changing them to powdery clay and dissolved salts. Other minerals more resistant to weathering withstood the chemical attack, only to tumble loose from the rock when the grains surrounding them became dust. Swift mountain streams caught up this mixture of debris and carried it down the mountainsides.

In the early mid-Cretaceous, while the sea yet lay some distance east of the highlands, rivers probably carried their loads of rock debris to near the foot of the hills, to drop much of it at first on the bedrock plain (that bedrock plain which was itself the eroded remnant of a still older mountain range[2]) as vast fan-shaped sheets of alluvial sand spreading out like great aprons upon the mountains' laps. Later the sea encroached farther upon the land. Rivers began to release their loads as river deltas into the sea itself. It was on the rich alluvial soil of these deltas that the prolific lignitic forests spread.

Not only on the site of the Vineyard, but along the inner portion of the entire Cretaceous continental shelf of North America such sediments thickened, fed by that ancient belt of highlands.

Finally the waters rose so that the swamps were overcome by salt water. The last trees died. The sea passed completely over the drowned plants, lapping inland finally to break against a shore far to the north and northwest. At times during the Cretaceous, fifty percent of the United States was submerged. Clouds of fine mud, spreading into the sea, marked the mouths of the rivers. Into the Vineyard region this mud drifted, suspended, away from the shoreline of that time, to settle down onto what had become sea floor. One of the marine creatures on this long-ago sea floor left the clam-like shell which, buried within the uppermost of the Gay Head clay beds, has revealed the marine nature of the clay.

Hundreds of feet of this mud accumulated. Its weight exerted

[2] See page 17.

tremendous pressure on the smothered plant remains of the buried swamps. It pressed down on dead vegetation, squeezing water and other fluids from the plants as from wet sponges. The peat of the swamps turned into a form of more concentrated carbon, or lignite.

One hundred and fifty feet of Cretaceous clay still can be found on the Vineyard, although not in any one place. To determine the total clay thickness, geologists have examined scattered occurrences all over the island, and have reasoned, by their relationships to other sediments, how the layers would fall into place one above another in complete chronological sequence.

This total thickness of clay is about twenty feet greater than the height of the high cliffs at Cape Cod Light, North Truro, and is nearly equal to the height of Gay Head's cliffs. Yet even this is merely a fraction of the original thickness of the clayey layers. Much has been plowed away by subsequent ice sheets, and much lies beneath the sea. It may be that Cretaceous deposits extend all the way down to bedrock, some 770 feet deep off Gay Head.

In fact, more than a hundred miles east of the Vineyard, on submerged Georges Bank, there are Cretaceous beds nearly 1700 feet thick. These beds, unlike those on the Vineyard, remained relatively intact while glaciers were active to their west. So undisturbed are they that they contain many rather well-preserved fossil remains of molluscs, crabs, starfish, even worms which lived on the Cretaceous sea floor.

A Span of 25,000,000 Years

The Cretaceous began about 120,000,000 years ago and ended about 70,000,000 years ago. The final 25,000,000 years of the Cretaceous produced the clay-lignite deposits, the earliest readable pages in the Cape-Island past. Thus, the visible history of Marthas Vineyard dates back some 95,000,000 years.

Let us digress a bit here, and try to gain some feeling for a time span of 25,000,000 years. Let us consider what was produced in another part of the world during that same time—and with the slowest of production methods and the most insignificant of raw materials.

About the time that the lignite plants were growing in southeastern New England, much of Old England and its neighbors lay engulfed by shallow Cretaceous seas. In these waters, as in today's seas, lived one-celled marine creatures, *Globigerina*, much smaller than the first letter of their name here. As they died, tiny shells sifted slowly to the sea floor—a fall which still continues at the same rate, but which accumulates so slowly on the bottom that no one in a lifetime could perceive the gain. Yet 25,000,000 years allowed two thousand feet of these shells to accumulate, each miniature skeleton becoming a tiny building stone in the massive chalk cliffs of Dover. Explorer Gosnold sailed by Gay Head and named it Dover Cliffs, little realizing what close geological cousins the two headlands are.

For Cannons and Collectors

At Gay Head the Cretaceous clay is most dramatic, like a bulkhead, despite the landslumps which partly cover it toward the bottom. And similar clays and lignites reach inland in undulating layers under much of the island (*Figure 3*). Although particularly

Figure 3. *An exposure of Cretaceous clay near Peaked Hills, Marthas Vineyard.*
U. S. GEOLOGICAL SURVEY

common up-island, Cretaceous lignite has been found even at Chappaquiddick, all the way across the island.

Color distinguishes the clay in places. Along Menemsha Pond it is red, as it is also high in the cliffs on the northwest shore of the island. Northeast of Makoniky it is blue-gray. The Nashaquitsa Cliffs near Gay Head have some grayish clay too, with lignite occasionally visible beneath it.

Such Cretaceous deposits on the Vineyard contributed handsomely to local economy at one time. During the nineteenth century, good-sized pockets of lignite showed up at Gay Head. Early inhabitants mined it, first cutting out blocks to warm their own hearths, later shipping it commercially to Salem for the extraction of alum. But alum could be bought cheaper in Europe and the Vineyard business finally failed.

The pristine white Cretaceous clay, such as that in some cliffs and beneath the surface near Menemsha Pond, is nearly pure kaolin (hydrous aluminum silicate), the sea-refined product of the weathering of highland rocks. It gets even whiter when burned, and the few impurities in it disappear. White-walled, water-filled clay quarries across the western part of the island indicate that formerly this kaolin was mined in large quantities. It once found city markets for such diverse uses as cannon molds, pottery, sugar refining and firebricks. By 1888, Boston ships were a common sight half a mile off Gay Head, at anchor to avoid the shoals, waiting for ferries to bring them loads of this clay. In 1893 a short-lived "Gay Head Clay Company" was organized, and for a few years its wharf, reaching far into the water from the beach near Lobsterville, thudded with activity.

The nonwhite clays are stained their various colors by mineral impurities. These have found uses too. Vineyard Indians used them for their colorful pottery. When settlers came and found an abundant supply available in the cliffs, they dug them and built roads. They crushed them, stirred them into a liquid base, and painted their houses with them. The brick-red clays near Menemsha and Makoniky became the bases of a terra-cotta industry. And Makoniky's light blue-gray clay became yellow bricks in the heat of the bake ovens.

Finally, with more of an aesthetic than a commercial value, mineral specimens occur in the Vineyard Cretaceous. They lie embedded mainly in the lignite, their crystalline beauty half hidden by organic material. They formed from chemicals produced by the incomplete decay of the ancient plants.

There are clear, glassy needles of gypsum (hydrous calcium sulfate). Marcasite, an iron sulfide, forms rusty-coated gray or grayish-bronze balls, some with beautifully radiating structures. Pyrite, another iron sulfide, appears like tiny brass cauliflowers or cubes. In fact, pyrite nodules seem still to be forming in alignment with the layering of the sediments. These two iron sulfides in places make up as much as ten percent by weight of the lignite. In addition, some brown pitchy siderite, a form of iron carbonate, has been found.

The Cretaceous on Nantucket and Cape Cod

Theoretically at least, Nantucket Island and Cape Cod, as parts of the coastal plain, should have Cretaceous sediments beneath their surfaces too. It has been suggested that the old blue clay with which Nantucketers once made cisterns for whale oil is Cretaceous in age. However, none shows up recognizably at the surface or in any of the wave-worn cliffs.

Thus, looking briefly backward over the last long 25,000,000 years of the Cretaceous, we see the slow growth of the Cape-Island frameworks on the rocky foundations. We see stands of luxuriant forests; then their swamping and eventual drowning by rising seas. Next we watch the long accumulation of clay on the sea bottom, which once had been forest floor and which one day in the far future would be partly exposed at Gay Head. We note the consequent arrival of sea life; and the refining of lignite and chemical creation of crystals from the thick plant remains. Now the lower layer of the framework is complete. As the Cretaceous sea withdraws and lays bare its clayey floor, we turn to the next stage of framework building, the Tertiary, to see what it will bring.

A New Era—The Tertiary

The slow withdrawal of the New England Cretaceous sea about seventy million years ago was part of a worldwide fall in sea level. During the death throes of the Cretaceous, tremendous forces which long had been acting on the crust of the earth in western America were causing it finally to yield in the series of slow bulges and upheavals that culminated in the Rocky Mountains. Similar forces pushed mountains in the eastern part of the continent to new heights.

The emergence of the drowned lands and the mountain uplift essentially marked the end of the Cretaceous Period and of the Mesozoic Era. This great earth revolution, like the political revolutions of men, was a time of trial, the end of one way of life and the beginning of another. Many creatures, thrown out of balance with their environment, faded into extinction. The ponderous dinosaurs disappeared forever. Mammals, heretofore the meek in the presence of those giant ruling reptiles, came into their inheritance. There began a new Era, the Cenozoic, and a new Period, the Tertiary.

The new mountains which rose with the Tertiary were vulnerable to the same erosive forces that had turned their predecessors to flattened clay and sand. A long period of erosion began. During this time, not only the weathering hides of the mountains, but also many of the earlier Cretaceous sediments, crumbled away. Rivers carried all to a retreating sea.

Cape Cod's Tertiary Swamps

During the early Tertiary (known as the Eocene Epoch), these rivers apparently meandered across the present site of Cape Cod. They crossed a flat, swampy forest lowland much like that which had existed in the Vineyard region during the Cretaceous. Today that forest lies eighty-six to some two hundred and sixty-four feet beneath the Cape. All that remains is lignite with microscopic plant spores and grains of pollen. These deposits were unearthed by geologists in 1960, when drills found the lignite beneath Provincetown. It was an exciting discovery, for they are the only early

Tertiary deposits known to exist in the region. At this writing, investigations are continuing.

The Vineyard's Greensand

In the middle of the Tertiary, some thirty million years ago, the land again sank and the sea again flowed over it. In this Miocene Epoch, not only the Cape Cod region, but even low areas of eastern Massachusetts and Rhode Island lay beneath shallow water. Where Marthas Vineyard rises now, seawater probably stood some sixty feet deep.

Near Plymouth on the mainland, wells have encountered thick fossil-filled deposits from this Tertiary sea. Far to the east of the Islands, deep beneath the surface of Georges Bank, marine geologists have found similar Tertiary beds. Thus, there must have been an extensive sheet of such deposits once along New England's coast. Today all that remain visible at the surface are the Tertiary greensands of Marthas Vineyard, where glaciers have shoved them high enough to be seen above sea level.

This greensand deposit, like the Cretaceous deposits preceding it, has been mutilated and wrinkled by ice. At Gay Head we can see it streaking prominently across the cliffs. The layer is greenish in some places, especially the lower parts, and reddish-brown elsewhere. The colors come from granules of the mineral glauconite. Glauconite is green when freshly exposed. Sands composed of it are known as greensands. It oxidizes to a reddish color on exposure to air and rainwater. Thus this Tertiary bed at Gay Head, red though it frequently is, is a greensand.

The Vineyard greensand ranges in thickness from a yard to ten feet. It is a sort of loosely cemented sandstone, consisting mostly of round, sand-sized grains of glauconite mixed with white grains of quartz. A closer look at the glauconite shows that many grains are made of tiny rounded fossils stained on the surfaces.

Larger fossils are common here too. It has been said that there are more fossils here than at any other part of the New England coast. There are blackened crab remains with perfect detail even to interior gills. There are bones and teeth of sharks and other fishes and of sea mammals. Many of the fossils are seashells such as snails and basket clams. Some of the clamshells are still closed and, with the crabs, lie embedded in the sand in the positions which the living animals would have taken on the sea floor.

Glauconite formed commonly in shallow waters of Cretaceous and Tertiary seas. It still forms today, apparently, at depths of from sixty feet to half a mile. Since it forms today only in seawaters, we can assume that the Tertiary beds of the Vineyard originated beneath the sea, especially since associated fossils are forms of sea life.

Some of the glauconite apparently was swept into the Vineyard area from elsewhere, carried in currents activated by storms; for numerous crab shells appear to have been rolled a long distance and worn smooth by water.

However, most of the glauconite probably formed chemically, through some long-ago chain of chemical reactions on the sea floor. During Tertiary times, some deposited itself about nuclei made of shells of one-celled sea animals related to the chalk-forming Globigerina. Some replaced other minerals or filled seashells.

New England's Mid-Tertiary Life and Climate

The fossil assortment tells us that the mid-Tertiary climate in New England was warm and pleasant. Sea animals found at present near the West Indies flourished then in these northern waters. Other little creatures would be familiar to Cape-Island beachcombers today. Crabs scuttled across the sand of the Tertiary sea floor; scallops zigzagged through the water; and in the mud nestled nut clams, mussels, quahogs, long-necks, Yoldias, cockles and razor clams.

Sometimes there were sudden influxes of new sand and mud onto the tidal flats, carried by flooding rivers. Such debris entombed these molluscs and crabs in their burrows beneath the surface and preserved them in life positions, making the greensand a sort of sea creatures' Pompeii.

There are signs of other remote and violent events also buried in the sands. Fossil teeth and bones indicate that sharks cruised the Tertiary waters, voracious threats to the young of the abundant sea mammals—whales, seals and walruses—which found their way into these waters as harbor seals still do. Schools of whales occasionally must have become stranded when they chased squid and small fish toward shore; times have not changed at all in this respect.

Most interesting of all, perhaps, is the sort of fossil sometimes found with the seal bones. It looks like a plain gray or black pebble, an inch or two across and polished as smooth and shiny as a tumbled gemstone. It acquired its polish in the stomach of a Tertiary seal, according to theory, and is called a gastrolith. Some present-day seals swallow stones, sometimes as large as goose eggs. According to Dr. John A. Moore of The American Museum of Natural History, certain species eat stones just before fasting, perhaps to reduce hunger pangs; others may carry stones to wage war on stomach worms. The Tertiary seals were wanderers; one gastrolith found in the greensand is of a type of flint known to occur naturally near Hudson Bay.

While all this life was thriving offshore, swamps flourished near shore, created by minor fluctuations in the level of the Tertiary sea. The climate probably was warm; at least one primitive mud-loving rhinoceros left a tooth in the greensand. Even a crocodile's tooth has turned up in this prolific bed, but it was silicified and apparently had come from elsewhere with some of the quartz of the greensand.

Upon the death and decay of sea creatures, a local abundance of phosphorus developed from the remains. This, plus incomplete decomposition, sometimes led to the formation of nodules of phosphatic decay products. Similar phosphate nodules lie strewn upon the bottoms of many offshore regions today. Electrical pulls lump these nodules around remains of crabs or seashells as dark, irregular masses an inch or two across. Risen from the sea with the Vineyard greensand, the Tertiary phosphate lumps lie embedded in it still. Even today, when rubbed together they may give forth a trace of disagreeable decay odor. Containing fifty percent calcium phosphate, they would make excellent fertilizer if they were in more concentrated patches. We may be sure that local farmers would have enriched their fields with them as they have done with the potash-rich greensand itself.

Nantucket's Middle Tertiary

Mid-Tertiary deposits may make up part of Nantucket's foundations, perhaps forming the clay which comprises the deep cores of some of the glacial hills, such as those along the eastern shore and in the western part of Nantucket Town. Clumps of greensand lie in the glacial sediments on Nantucket, which may have come from Nantucket itself or from north of it, scraped up from the wide sheet of Tertiary sediment on the sea floor. But the island can boast at least one ancient swamp deposit thought to be Tertiary. In it are embedded drops of fine, pale, wine-colored amber, up to a pound in size. Within the amber have been found a tree leaf, a fly and some ants which had the long-ago misfortune of entangling themselves in the sticky gum.

Glancing backward across the Tertiary scene on the Cape and Islands, we see first the withdrawal of the Cretaceous sea, coinciding with the rise of great mountains on the mainland. In the Cape Cod region, perhaps on Nantucket too, we see forests swamped and finally drowned by rising seas; and the subsequent formation of lignite. Across the Vineyard and Nantucket region we see next a new sea rise; a sheet of greensand spreads across the shallow sea floor, and we see in the water and on the sea bottom a rich assortment of sea life which is to leave for us its fossil record of bones, teeth and shells.

Stumbling Blocks for the Glaciers

Then the Tertiary curtain fell and the seas withdrew. Exposed to sun, air and erosion lay the newly created coastal plain. Beneath its gently sloping surface were more than one thousand vertical feet of soft Cretaceous and Tertiary sediments. This lowland plain stretched along the entire Atlantic seaboard. Today it still makes up the seacoast south of New England, and many parts of it are nearly as featureless as when they formed the floor of shallow seas.

However, near the coast, many roads which run at right angles to the shore climb up and down over low hills. These hills stretch in wave-like ripples parallel to the shore and are separated by flat, valley-like lowlands. Erosion carved this pattern in the coastal-plain surface during the long interval between the withdrawal of Tertiary seas and the coming of the glaciers. It happened in the New England area as well as farther south. But in New England, later glaciers were to leave their marks. Pressing down with massive weights, they would cause the land itself to sink. And when their ice finally melted, its water raised the sea. Flooding waters filled the coastal-plain valleys and beat against rocky inland hills, such as those which today form the Maine coast. A few coastal-plain hills were left covered with such thick ice debris that their tops have remained visible even above the postglacial sea level.

Cape Cod, Marthas Vineyard and Nantucket are such visible hilltops, the final remnants of a once extensive coastal plain. The buried backbones of the Cape and Islands are the old coastal-plain hills. For on its travels south across New England, the ice, loaded with incalculable tons of earth waste, stumbled over those hills which lay across its path. Thus slowed up, and made sluggish as well by the warmer latitude which it encountered here, the ice dumped its load on these hills. The piles of glacial debris on the ridges—the flesh on the skeletons—are what give us today's Cape and Islands.

Cape Cod's buried ridge extends roughly parallel to the Connecticut-Rhode Island seacoast, then swings northward to parallel the Massachusetts-New Hampshire seacoast. The Marthas Vineyard-Nantucket ridge lies south of and parallel to the Cape ridge. Between these ridges is the valley which now holds the waters of Vineyard and Nantucket Sounds. North of the Cape ridge is the valley which makes up the basin of the Gulf of Maine. Enclosed in the curve of the Cape ridge is the basin of Cape Cod Bay. Southeast of the Cape and Islands, the submerged seaward extensions of the Island ridge are the foundations for some of the great fishing banks of New England: the Nantucket Shoals, which lie a few miles southeast of Nantucket on the gently sloping back of the Island ridge; and Georges Bank, some ninety miles farther seaward.

During the final stages of hill and valley carving, earth time flowed into a new Period, the Quaternary. This is divided into two Epochs, the Pleistocene and today's Recent. Both man and the major part of the Cape and Islands first made their appearances during the Pleistocene Epoch.

The Vineyard's Bony Gravel

The early Pleistocene apparently has left no record on the Cape or Islands. The first Pleistocene deposit which we find in this region is a gravelly layer on Marthas Vineyard, twelve to eighteen inches thick. It is thought to date back to the second half of the Pleistocene. It was the last known bed to be deposited in the Cape-Island region before the coming of the Pleistocene ice sheets.

This Aquinnah Conglomerate—named after the Indian name for Gay Head—is a rough natural concrete. It appears directly below Gay Head's lighthouse above the white Cretaceous clay. A hodgepodge mixture of white quartz nuggets, black chert and colored chalcedony, sand fragments, and pieces of shell, it lies loosely cemented with more quartz and tan iron oxide.

Scattered throughout the deposit are bones and teeth, so abundantly that an early name for the bed was "osseous conglomerate." There are teeth, ribs, jawbones, paddle bones and skull bones of whales and porpoises. Even in recent years these ancient bones, blackened by seawater, sometimes have lined the strands. Not uncommon are three- to four-inch shark teeth. An early geologist estimated that any shark with teeth that large must have been more than sixty feet long, with a mouth that gaped open sufficiently to encircle the tallest man.

These bony fossils doubtless came from sea creatures which died and left their remains just offshore while rivers were pouring loads of pebbles into the sea. But among these pebbles are other fossils, picked up by streams inland and carried as part of their alluvium. For instance, many of the shell fragments are of this sort; they were fossils even during the Pleistocene. These corals, crinoid (sea lily) stems and graptolites lived perhaps three million years ago, during the remote earth Period known as early Devonian. During the long, long interval between Devonian and Pleistocene, their remains became replaced with chert and, picked up with other chert by eroding rivers, they became incongruous parts of a Pleistocene deposit.

The upper part of the Aquinnah Conglomerate differs a little from the older, lower part. In the upper part there are fossils recognizably derived from Tertiary greensands. And the deposit once yielded the bones of a land mammal, a camel, and above that, toward the very top of the conglomerate, part of the skeleton of a wild horse.

From all these clues it is not difficult to reconstruct events. The large size of the pebble grains tells us that there must have been active erosion of the land during that part of the Pleistocene. Perhaps the sea did not even cover the Vineyard area then; more likely the Aquinnah is the work of shallow rivers which poured rapidly onto the low-lying coastal plains, abrading these materials and washing from their banks the quartz and chert. These streams reached the edge of the sea where the Vineyard now is. They deposited their alluvium, including the cherty fossils which had been torn from far older strata to the west.

After the influx of pebbly debris, erosion took over, attacking the newly deposited gravel beds. Coastal-plain sediments, including Tertiary greensands and Cretaceous clay, also were vulnerable. Fragments from these materials, including greensand fossils, became new deposits. Reworked by the rapid rivers which channeled the land, these new deposits piled up on the coarse conglomerate which forms the lower part of the Aquinnah.

Then, while the sea yet lingered at its new low level, life arrived on the just-emerged coastal plain. Wide grasslands must have spread over the lowlands, thriving in the warm, arid climate. And wild camels roamed the plains. . . .

The earth weaves its creations slowly, doing and undoing, tearing apart at night what was created during the day, endlessly working, never finishing. For centuries there was only erosion. Then the sea rose once more, but only moderately. Enough, however, for its currents to fill the valleys of the coastal lands with sand and clay and to leave the higher of the coastal-plain hills as small, low islands —the not-very-impressive net results of nearly 100,000,000 years of work.

CHAPTER II

Sudden Winter on the Globe

> *The long summer was over. For ages a tropical climate had prevailed over a great part of the earth, and animals whose home is now beneath the Equator roamed over the world from the far South to the very borders of the Arctic. . . . Then a sudden intense winter, that was also to last for ages, fell over the globe. . . . If the glacial theory be true, a great mass of ice, of which the present glaciers are but the remnants, formerly spread over the whole northern hemisphere, and has gradually disappeared. . . . Every terminal moraine is the retreating footprint of some glacier, as it slowly yielded possession of the plain, and betook itself to the mountains; wherever we find one of these ancient semi-circular walls of unusual size, there we may be sure the glacier resolutely set its icy foot, disputing the ground inch by inch, while heat and cold strove for the mastery.*
>
> LOUIS AGASSIZ
> *Geological Sketches*

As we look at the upper part of the Aquinnah Conglomerate at Gay Head, scarcely visible is the zone of division which separates the layer where the fossil horse bone was found from the part below containing the camel bone. Yet in the history of the Cape and Islands it is a significant division.

Let us allow the mute evidence of these two layers to lead us back in imagination through the millennia. We find ourselves standing on a wide grassy plain. We are in the period during which the materials of the lower of the two layers are being laid down. Over this plain roam small early camels, no strangers to the United States countryside at that time.

Now let us leave and permit the slow processes of the earth to add new layers to this plain, layers of gravel and debris discarded by streams as they near the ends of their journeys. Many centuries later, let us return.

Again we are on a plain, but its surface has been built up higher than the former one. We feel ever-increasing winds sweeping the land, and the warmth of the sun is lost in more and more chilling fogs. The first faint breath of growing glaciers, forming like gathering armies in the north, is on the land. We leave and allow a few more centuries to pass. When we return we are still a few inches higher in the conglomerate beds. No longer do we see camels; instead, wild horses graze on the grasses, for horses are more suited to the newly cool, moist climate. Even as we watch these horses roam, huge sheets of ice have begun their slow journeys from the north, eventually to bury the land on which we stand beneath millions of tons of ice and debris.

Meanwhile, with the chilled air drifting across the land as if from a gargantuan opened refrigerator, the temperature of the neighboring seas is dropping gradually, perhaps as much as eighteen degrees. A cold-water fauna slowly is replacing the temperate climate species that had lived in these waters. By this time we find that we have come to the top of the Aquinnah Conglomerate. Quickly let us return to the safety of the present. For at this point the ice itself arrives.

The Ice Ages

The Pleistocene Epoch has excited popular interest and imagination as no other geologic division. It alone has earned a popular name, the Ice Age. Actually, during the Pleistocene there were four distinct glacial stages, during each of which nearly a third of the face of the earth possibly lay smothered by ice.

Each of these periods of glacial conquest lasted only twenty or thirty thousand years. The interglacial stages which separated them were many times that long. During each interglacial stage, warming temperatures melted the glaciers back until they nearly disappeared, and the land which they had covered became clothed again in greenery and populated with animal life, even as today.

Evidence indicates that the Ice Age story on the Cape and Islands began immediately before the fourth of the glacial stages, known as the Wisconsin Stage.

With the waning of the last ice sheet a mere ten thousand years ago, the Pleistocene Epoch came to an official close and the earth passed into today's Recent Epoch. Man has dominated the Recent, yet all of his recorded history is encompassed in less than half of it. And the total of Recent time would amount to only a fraction of one of the interglacial stages of the Pleistocene.

The Pleistocene has ended, but have the Ice Ages? We cannot be sure. It is not unlikely that we are living in an interglacial stage. If this is so, and if the powerful ice sheets should be reborn and reach the size of the former glaciers, they would bring a wave of property destruction for a length of time and on a scale which would make the ravages of all the wars of history seem insignificant. Slowly the ice would nudge populations of the earth's higher latitudes away from their homes, crowding them together in ever-narrowing temperate zones. Many of the world's great cities would be as sand castles built on the shore to enjoy a fleeting existence between advances of the tide.

There were, of course, no cities to fall before the successive tides of Pleistocene ice, but there were men who waited out the more recent of the bitter glacial climates in the caves of southern Europe, Asia and North America. In Spain and southern France, ice drove reindeer and mammoths—tundra animals of the north—to the front yards of Cro-Magnon man, who hunted and ate these creatures, recording their existence with graceful artistry on the walls and ceilings of his caves.

Wherever the ice went it rearranged things on the earth like a dissatisfied housewife. It swept away much of the rich, age-old accumulation of soils which covered south-central Canada, and thus uncovered the immensely valuable iron and copper ores in the bedrock. It hauled the stolen soils across the border, dumping them as a thick blanket over the north-central United States to create her amber fields of grain. It rearranged the shapes and drainage of the Great Lakes, scooping them into the forms they have today and providing the opportunity for the recently built Saint Lawrence Seaway. Indirectly, the ice even rearranged life on the earth. Nearly eight million cubic miles of ice imprisoned the earth's waters. This resulted in a drop in sea level which opened up land bridges between continents now separated by water. One of these was the present Bering Strait. Glaciers did not cover this region, and across it there probably occurred a two-way traffic of life which brought man to North America.

Still like an industrious housewife, the ice finished its rearrangement with a redecoration of the earth's surface. It molded the soft contours of the north-central farmlands and polished the rocky walls of New England and Scandinavia. It gave to Boston its Beacon and Bunker Hills and to Minnesota its ten thousand lakes. Finally, in several happy spurts of activity, it built up most of the bulk of Cape Cod, Marthas Vineyard and Nantucket.

Glacial Birth and Movement

Glaciers are conceived in high latitudes or altitudes where a cold, wet climate produces large quantities of snow; they are born when continued cold allows little to melt from year to year. In a glacier we see the snows of many yesteryears.

Glaciers begin as snowfields. When the climate remains right for them, they grow and thicken, snow upon snow, blizzard after blizzard. At last they reach one or two hundred feet. As they approach this thickness, the whole heavy mass presses down on the oldest snows at the bottom, compressing the delicate, six-rayed flakes. These recrystallize into compact, mosaically interlocked ice crystals. Meltwater from the surface trickles down and refreezes, cementing individual grains still more firmly together. As the weight of additional snow becomes more than lower snow can withstand even by recrystallization, the mass begins to flow outward in all directions. When such self-produced movement begins, the glacier starts its existence, with a life expectancy dependent on the climate.

The familiar minor glaciers cradled in valleys flanking high mountains of the world today owe their existence to weather conditions peculiar to and partly caused by mountains—cool temperatures and abundant precipitation. The journeys of most of these are short and unspectacular, dwindling to an end when the glaciers reach the foot of their mountains, if not earlier.

Glaciers of the Pleistocene, however, nourished by a series of crucial global climate shifts, grew and merged into enormous continental ice caps. The great ice blanket covering Greenland today, eleven thousand feet thick in the center, is only a lingering remnant of one of these. Such continental glaciers, unlike today's mountain glaciers, did not merely fill mountain valleys; in their ascendency they flowed over entire mountain ranges that happened to stand in their paths.

At the beginning of each of the four Pleistocene glacial stages, glacial growth began in scattered areas of snow accumulation dotted through the earth's higher latitudes. From the barren uplands of Labrador and the lofty peaks of the Canadian coast ranges; over the rocky highlands of Scandinavia and Siberia and, a world away, over the mountains of Patagonia and southern New Zealand, the great ice sheets began their grinding journeys downhill, moving four or five feet a day.

Along their thinner advancing margins they tended to split around obstacles, forking apart like a wave which encounters a rock near the tide line. Later the thick central part of the ice might ride over the obstacle, but the forward margin was apt to retain its irregular scalloped shape as it continued to advance.

As the glaciers moved they incorporated into themselves whatever materials were not held tightly enough in the grip of the earth. They picked up soils and parts of frozen forests, sand, mud, clay, pebbles and boulders. They pried loose rocks the size of houses from their native ledges. As this material accumulated and was processed in the mill of moving ice, pebbles and cobbles tended to crumble to sand and fine powder (rock flour). Larger, tougher boulders survived, although scored and scratched, to be dropped as erratics many miles, perhaps, from their places of origin.

Along its route the ice lost much of its load as ground moraine, where friction against the land surface tore loose debris from the bottom of the ice and left it veneering the ground after the ice had melted and gone. The rest of the load ended as thick ridges, festoon-like frontal moraines, wherever a warming climate melted the ice margin back as fast as the main mass could push forward (*Figure 4*). Although in such a place the glacier would appear to be standing still, the body of moving ice actually would bring a continual supply of new debris to the front, where the front edge, melting, would drop it.

Whenever the sun warmed the ice surface, especially during summers, streams of meltwater flowed across the top of the glacier, down its front and underneath it. Naturally these carried with them some of the glacial debris. Only the swiftest of meltwater streams, born during times of rapid melting, had the strength to move large boulders, although huge rocks sometimes traveled on

Figure 4. *A glacier building a moraine. This might have been the scene at any Cape or Island moraine some forty thousand years ago.*

U. S. GEOLOGICAL SURVEY

floating ice rafts in these streams. Most of the meltwaters' loads consisted of pebbles, gravel, sand, mud and clay. The speed of the running water sorted these neatly into graded sizes.

Fans of such glacio-alluvial debris grew, often spreading many miles from the melting ice sheet. When the ice reached a temporary or final stopping place, of course, meltwaters were particularly lively. Therefore, while a moraine piled up at the edge of the ice, outwash deposits might spread out beyond the moraine and might reach even farther than the farthest range of the ice itself. During vigorous melting, the fans of debris laid down by individual streams spread and joined into thick outwash plains.

All of the debris of frontal or ground moraines is called till or drift, although drift, strictly speaking, is a more general term, which can include both till and outwash.[3]

The materials carried in glaciers provide two important clues by which former routes of the ice may be reconstructed. Most rocks have composition-texture combinations that are distinctive. Many large pebbles and boulders scattered through till carry such a seal of identity. Like the band placed on the leg of a migrant bird, it provides the finder with a means of retracing the route of migration.

Furthermore, the concentration of abrasive materials in the lower part of a moving glacier, coupled with an enormous downward pressure of thousands of tons, transformed the glacier into a giant sander which scratched deep grooves in the solid rock over which

[3] The origin of the term, drift, involves an interesting glimpse at the youthful days of geology. Prior to the 1840s and the work of Louis Agassiz, one of the founders of the glacial theory, there had been no apparent explanation for the thick, widespread deposits of strange clays and boulders across northern Europe and the northern United States. Many rocks were perched in odd places where no known geologic process might have put them—except, as Hitchcock first theorized, "some powerful current of water which, in early times, swept over the globe." Some maintained that current to have been the Biblical Flood. Accordingly, these deposits were termed diluvium, drifted deposits, or drift. That is how Hitchcock's preliminary report in 1833 classified Cape-Island deposits—although the author was dissatisfied with the lack of a reasonable explanation for the source of such drift. But when Agassiz' theory came out eight years later, Hitchcock was quick to recognize the possible glacial nature of these materials. However, the immensity of the processes which we know now to have been involved are difficult to imagine, and Hitchcock still had reservations about the ability of any ice sheet to carry so much debris such a distance from the mainland. Referring to his tentative adoption of the glacial theory with respect to the "diluvial" phenomena on the Cape and Islands, he said in an address in 1841:

". . . Another difficulty [in the recently formulated glacial theory] results from the fact that some of the most remarkable of our moraines are found not in valleys, but on the seacoast, some of them fifty and others one hundred miles distant from any mountain much higher than themselves. I refer to those remarkable conical and oblong tumuli of drift, sometimes more than two hundred feet high, which occur in Plymouth and Barnstable Counties in Massachusetts. . . . I do not mention these difficulties (to which I might add more) as any strong evidence against this theory. For so remarkably does it solve most of the phenomena of diluvial action that I am constrained to believe its fundamental principles to be founded in truth. . . ."

it passed. After the ice disappeared, these striations, aligned in the direction of movement, were left to point out the glacier's path. Many rock surfaces of New England are covered with them: you can follow them like signposts, for instance, on the bare rocks of Maine's Acadia National Park. Across the rest of Maine too, and New Hampshire, Massachusetts and Rhode Island, these scratches mark out the routes of the Pleistocene ice as it gathered up the varied materials with which it would create Cape Cod, Marthas Vineyard and Nantucket (*Figure 5*).

The Ice on North America

Wave-cut cliffs along the shores of the Cape and Islands reveal that they are built up like giant layer cakes, capped by a green frosting of vegetation. With the exception of the old twisted Cretaceous, Tertiary and early Pleistocene sediments at Gay Head, these layers tell the story of three successive waves of ice from the north. The mountainous moving lobes carried embedded in their bodies the entire blanket of soil which had covered New England and even countless tons of its rocky framework. Again and again these lobes pressed southward from Labrador, strewing parts of their burden beneath them as they moved, dumping most of it in piles at their final termini where they melted—ponderously building the Cape and Islands layer by massive layer.

Whenever the ice lobes rested and created a moraine, meltwater streams piled outwash plains before them. Whenever warming climates drove back the front margin of the ice, nonglacial marine beds were laid down in the rising seawaters which, fed by melting ice, encroached upon the land.

Each of the three glacial invasions was relatively short-termed, for all probably occurred within the Wisconsin glacial Stage. This, the last of the four Pleistocene glacial stages, started some thirty thousand years ago. During each period of ice retreat, the glaciers probably melted back not entirely, but only enough to lay bare much of New England.

Labrador was the birthplace of all the New England glacial lobes. Here, between desolate summits and high within intermon-

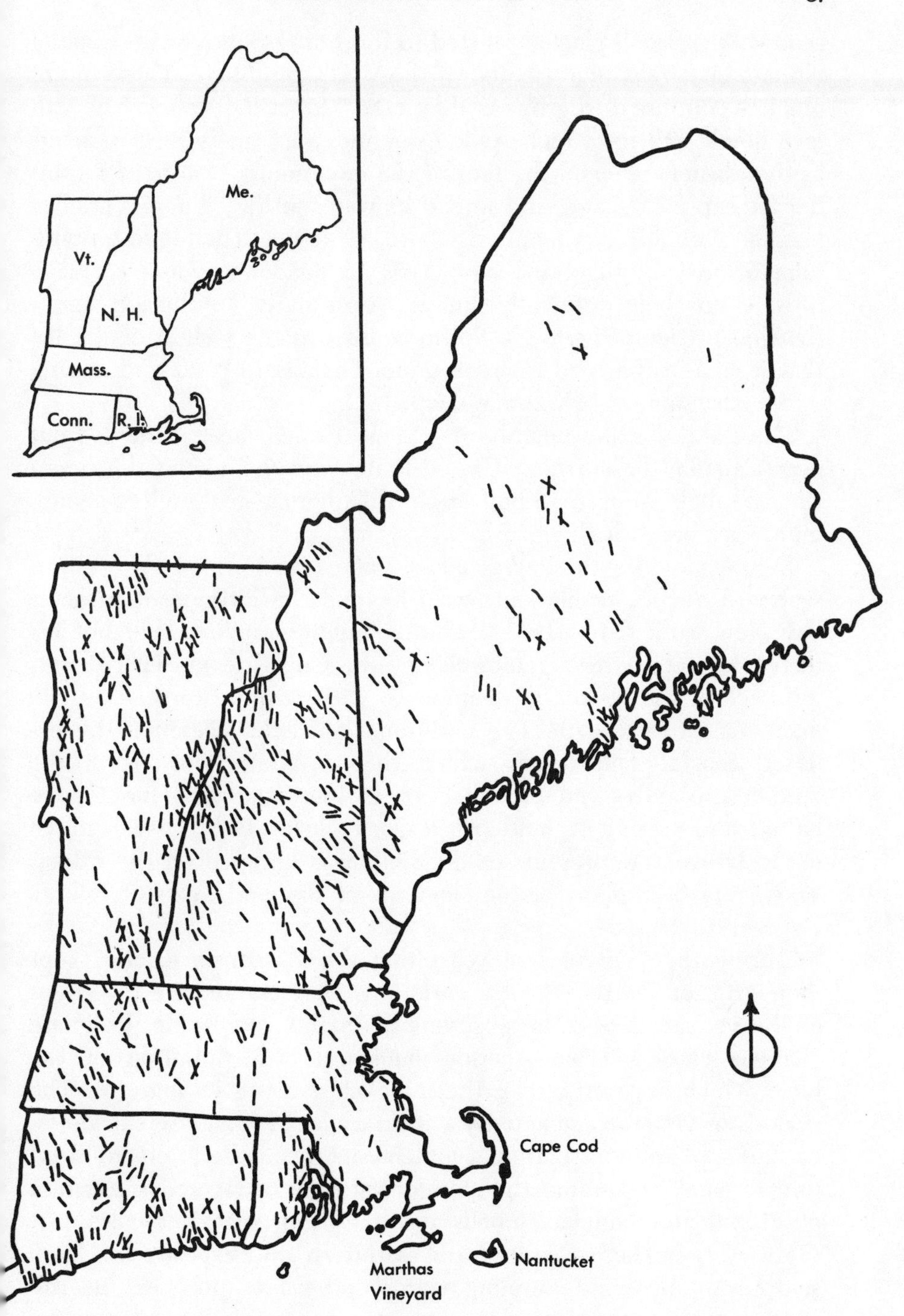

Figure 5. *The trail of the glaciers across New England. The map shows the general trends of rock scratches carved into New England's bedrock by the Pleistocene ice.*

tane valleys, small glaciers started to form in response to the cooling climate change which ushered in the Wisconsin Stage. These infant glaciers continued to grow as they crept down the slopes. The valleys filled with ice which spilled out into vast, merging ice sheets as the glaciers reached the foot of the mountains. The great Labrador ice cap took shape, and spread its way southward across eastern Canada with a frozen immensity hard to picture. The highest mountains of New England and New York are inadequate as measuring rods; even their mile-high summits eventually lay buried under the thickest ice (*Figure 6*). So immense was the weight of the ice that it caused the hard bedrock under Canada to yield and to sink into a basin almost half a mile deep.

The Labrador and neighboring Keewatin ice sheets sprawled out over Canada, the northern United States and the northeastern continental shelf. More than half of North America—five million square miles—lay beneath them.

In none of the three Wisconsin-Stage invasions on the eastern seaboard did ice reach southward beyond the latitude of present-day New York City. Here it abutted against an invisible but effective climate barrier, reinforced by the Gulf Stream, where melting back of the ice margin kept pace with pushing forward of the main mass to its north. The additional obstacles presented by exposed coastal-plain ridges—which then lay as extensions of the land or as islands—were too much for its thinning edges. Here the ice halted and loosed its hold on its rocky load. Debris was strewn not only over the present sea floor but was dumped on the ridges. Thus were born many of the elements of the landscapes of today's Cape and Islands.

During the thousands of years that the Wisconsin glaciers kept their grips on North America, variations were common in the edges of the ice sheets and the positions of their lobes. Cyclical climate changes caused a slow-motion back-and-forth pulsation of the lobes which fingered out of the main mass along its margins. The periods of vigorous ice activity within a glacial stage are substages, or stadia. Waning of marginal ice characterized the interstadial intervals which separated the stadia. Naturally a layered succession of glacial and nonglacial beds resulted from such fluctuations.

Every time the ice melted and withdrew and exposed the land, sculpturing by wind, running water and waves prepared its till-strewn surface for a landscape which the spread of vegetation

Figure 6. *A glacier flowing over a mountain range. This might have been the scene in New Hampshire's White Mountains during the building of the Cape and Islands.* U. S. GEOLOGICAL SURVEY

would complete. The hills and hollows carved into the land at this time would have their own minor influence over the path of the next ice lobe to creep forth over the same region. During these interstadial periods, winds and waters acquired debris from the till as a by-product of their sculptures. Much of this reached the sea and became ordinary nonglacial sediments which, because they lie between successive tills, are called intertills. The materials of each of these intertill beds would in turn become grist for the ice which followed, just as the first glacier had derived some of its till from preglacial sediments flooring the region over which it had passed. For despite all of her apparent extravagances and profligacy, the earth husbands her supplies frugally, and in the never-ending cycle of geological creation, each grain of sand may be used over and over again.

Each ice advance across New England announced its coming by turbulent and unpleasant conditions fringing its borders in a zone many miles wide. For a distance of a hundred miles south of the moving ice front the air was chilled. The land lay locked in frost throughout the year, swept by winds and storms and often cloaked in fog. Directly along the ice border, low-growing tundra heath and Arctic plants appeared and provided grazing for the woolly mammoths and the herds of caribou that ranged over this narrow zone. Farther from the ice were spruce and fir forests like those found in subarctic regions today. These were the homes of great mastodons and of moose, deer, bears, boars, and horses.

As the ice pushed forward, gathering its unimpressive speed of up to one hundred fifty feet per year, these life zones moved outward in front of it like a retreating army. The movement was slow indeed, since it was determined by the rate of ice advance. Certainly no individual animal fled in conscious retreat before the ice; rather, the shift was one of groups and populations, following the southward-moving vegetation zones on which they depended for food.

Many sea animals of the formerly temperate waters responded when water temperatures dropped by moving away from the ice. Arctic species moved in. Remains of various creatures in marine sediments laid down before, during and after glaciations are vastly more numerous than remains of land animals and are valuable evidence in spotting climate changes in southeastern New England during this period.

Along the New England coast the ice scraped across the then-exposed continental shelf. It even spread beyond the land over the shallow sea floor, terminating in an ice cliff along its eastern seaward edge. Here ice continuously broke off to send enormous icebergs thundering into the sea. Doubtless the sea itself in this region was covered by drifting sheets of pack ice, which met destruction when they reached the latitude of present-day New York and the warmth of the Gulf Stream.

Across the New England coastal plain and continental shelf the ice made several advances. During each of these it piled glacial debris around the Cretaceous and Tertiary cores of what would eventually be Cape Cod and the Islands. For the remainder of Part One we will look at each of these ice advances in turn. We will watch their ponderous progress and recessions, and follow the enormous processes by which they shaped the Cape and Islands.

Difficulties in Investigation

To figure out the correct sequence of the ice-laid materials which form most of the Cape and Islands is a jigsaw puzzle, complex and fascinating. Inland, one can, in many places, trace the path of an ice sheet right up to the ice deposits themselves by following scratches on rocks, but when we consider the Cape and Islands, such clues drop from sight beneath water dozens of miles from the moraines. Inland, bouldery tills left by a glacier usually differ so much from sandy outwash that one is easy to distinguish from the other, but on the Cape and Islands much of the till came from a former sea floor and in places looks just as sandy and boulderless as stream-carried outwash. Inland, one exposure of till can be followed right up to another miles away, and so they may be correlated, but with the Cape and Islands such relationships lie hidden beneath the sea. Finally, ice deformation in the Cape region was extreme; it kneaded Cretaceous and Tertiary sediments, early tills, outwash gravels and interstadial marine silts into geological marble cake.

A century of careful investigation has untangled much of the snarl, however—and with the strands of information we now have we can reweave this region's glacial history.

The First Ice Invasion—Marthas Vineyard

The first ice to invade the New England coastal plain probably came, as we have seen, during the early part of the Wisconsin Stage of the Pleistocene Epoch. It was an offshoot of the Labrador ice sheet which then lay sprawled across New England. This ice advance has been termed the Jameco substage of the Wisconsin Stage.

At Gay Head, Marthas Vineyard, about halfway up the face of the cliff above the Aquinnah Conglomerate there are two clear beds of loose glacial debris, weakly layered clay, sand, gravel and boulders. The boulders make these beds distinguishable from earlier nonglacial deposits.

The lower of the two glacial beds begins some fifty feet up the cliff face and is about fifteen feet thick. It is mainly sand and gravel. The entire bed has an ancient, weathered look, due to the rusty, dull and tarnished surfaces of its larger pebbles. This comes from the iron which gives color to the red clays in the drift layer above. From the clays it is leached by trickling waters and carried downward to stain the pebbles of the lower layer.

Forty years ago, where the iron-stained pebbles now appear, there was in the same cliff face a jumbled pile of much larger, nearly decayed boulders which seemed to make up the entire bed. It was named the Dukes Boulder Bed (after Dukes County). Some of those rocks reached eight feet across and were four tons in weight, and at least one was found which bore glacial scratches. The erosive forces which brought those rocks to light since have removed them, but not before they served geologists. For among the many common-type boulders in the rock pile there were some chunks of the rock, peridotite. This rock, black with magnetic iron ore, is one of the heaviest and darkest of rock types, a close but poor relative of the diamond-bearing rocks of South Africa. Gay Head's peridotite was ideal as a tracer of glaciers. It was relatively rare as a rock type. It was easily recognized and uncommon throughout all of the New England bedrock with the exception of one place in northern Rhode Island—Iron Mine Hill, east of Woonsocket. From this hill, only a thousand by six hundred feet in area and only sixty feet high, the glacier carried boulders to Marthas Vineyard and left them traceable back across those sixty miles. Today the large boulders are gone, but fragments of peridotite still appear among the drift pebbles. Thus we know that the finger of the pioneer ice which touched Marthas Vineyard passed over Woonsocket, Rhode Island, moving about S 40° E.

Meanwhile, a more easterly part of this same ice sheet reached the region where coastal-plain sediments overlapped part of New England's bedrock. Far to the east, where these loose sediments were still submerged, large quantities of them found their way into ice which scraped across the floor of the ocean. At that time the depth of the shallow coastal waters was enough only to carry them partway up the towering front of one of the ice lobes. However, there was also a wide expanse of coastal plain which, now submerged, was then land.

Indeed, twenty miles south of the Vineyard, beneath today's seawater, there is some peculiar sand; under the microscope its grains show the frosted and etched surfaces of ancient dune deposits, which may have formed in this region after the sea had ebbed away and before the ice had arrived. They may indicate that as glaciers approached, coastal-plain beds on parts of the present sea floor lay exposed to the air. The water which covers them today was then locked up within the ice.

Back at Gay Head, above the old Dukes Boulder Bed layer, of which little remains, there is a bed of stony blue clay, partly slumped into obscurity, called the Manetto Formation. It lies some seventy-five feet above the beach and its upper part is clay nearly filled with boulders. These are up to several feet across. Some of the larger rocks which stud the beach sand at the foot of the cliffs doubtless came from this layer. Beneath it, white gravel streaks through some twelve feet of sand. The gravel is very much like the preglacial Aquinnah Conglomerate, which actually lies much lower in the cliffs. Red clays are shuffled in with the white materials of this sandy layer, clays which are scarcely to be distinguished from the red clays of the Cretaceous on the Vineyard.

Apparently the first glacier did reshuffle the stack of beds at Gay Head. It scraped across the coastal-plain sea floor and slipped old Aquinnah Conglomerate and still older Cretaceous clay into higher, later glacial drift.

There is some particularly interesting outwash gravel at the eastern end of the two-mile, one hundred and fifty foot high Nashaquitsa Cliffs. Ice has forced it upward here to lie vertically beside Cretaceous clay at the base of the region known as the Weyquosque (or Wequobsque) Cliffs. The outwash formation is named the Weyquosque Formation. The pebbles of its upper part are colorful pieces of mainland granite. The debris beneath this granitic gravel contains many old friends from coastal-plain beds; we find Tertiary greensands, brown sandy silt, Pleistocene Aquinnah Conglomerate and many ancient sharks' teeth, scattered about like toy sailboats capsized in the turbulent swells of debris. Unfortunately, fallen material seldom allows us to see the lower part clearly.

At the western edge of Nashaquitsa we can spot more of such sandy outwash down near the level of the beach, but lately it has lacked the pockets of colored clay and fossils which would entitle it to claim coastal-plain sediments as ancestors. In fact, this outwash looks like ordinary beach gravel—quartz pebbles mixed with glittering paper-thin flakes of mica.

Perhaps further erosion of the cliffs will reveal fossils here too. Indeed, part of the unique fascination offered by this region's geology lies in expectations of new revelations to be raked into view at any time by waves, endlessly probing into the land.

Most of the Jameco-substage glacial drift so far encountered seems to be outwash rather than till; therefore none indicates with certainty where the ice actually lay. However, there is one place on the Vineyard where remnants of seemingly true till are exposed. This till is called the Moshup Till, honoring the legendary Indian giant Maushop, who resides in the Devils Den. Moshup Till is thrust up vertically at the Weyquosque Cliffs to lie beside the Weyquosque Outwash Formation, which actually was deposited first. We can spot the till by its large boulders, up to eight feet across. Some seasons, as we walk along the beach here, we must pick our way over scattered ramparts of these rocks; one section of the beach is known as Boulder Point. Low tide reveals that these rocks trail into the ocean like stepping stones of giants. Their seaward chain indicates by its extent the onetime reaches of the land itself, eaten back now to its present profile by millennia of winds and waves such as those which even at this moment are changing present to past.

The early ice which left the Moshup Till probably failed to cover all of the Vineyard, but at least we know that it did cut across the island's southwest end. Here, at the coastal-plain ridge and near the warm Gulf Stream, its vitality ebbed. It stopped, probably not far south of South Beach. It was the earliest-known ice to reach southeastern New England.

As we have seen, some of the Jameco deposits are bedded like outwash deposits, yet contain boulders here and there. We can conclude that rapid melting of surface ice in this frontier zone, particularly during summer seasons, must have produced torrential meltwater streams. These, murky with fine debris and noisy with the clatter of large boulders, coursed out from the ice margin. They left their loads strewn across the western part of the island as transient deltas and as sheets of alluvial sand and gravel.

No glacier can pass over a region without tampering with its contours. Even this early lobe, declining and hesitant as it may have been, was nevertheless a mighty rasp. As it moved from the northwest it left its mark on the ridged coastal plain; traveling nearly parallel to the lowlands, it gouged them deeper, carving the ridges into stronger relief. It reached the coastal-plain foundation of the Vineyard, moving at right angles to the island's present northwest shoreline, between Gay Head and Lambert's Cove. One tongue in advance of the larger lobe licked out into the area now occupied by Menemsha Pond, between Gay Head and Chilmark. The tongue thrust itself through the soft preglacial sediments, pressing them aside with a strength reinforced by the mass of the larger lobe behind it. As it spread to either side—to the southeast and the southwest—it shoved aside the earlier sands and gravels, forcing them into folds as the prow of a moving ship forces the water up into waves. Unlike the fleeting waves of water, however, the folds of these earth beds have remained, arching on either side away from the vanished ice.

Following this small tongue came the main lobe. It rode over the folded beds, tearing off some of the higher crests and leaving its coating of Moshup Till and waterborne gravels strewn across them and in the depressions—another layer in the building of the Vineyard.

From the relatively small pieces of evidence we find of the Jameco glacier, we can only roughly reconstruct part of its travels. We have seen that the ice, traveling from the Rhode Island region, covered the southwest end of Marthas Vineyard. From the ice front spread sheets of outwash, at least across the Nashaquitsa area. The ice plowed its way through the soft sediments around Menemsha Pond, and left Moshup Till on the land. How much farther it may have reached—in what other sorts of land molding it may have indulged—we can only speculate.

On Marthas Vineyard, in any case, we may see not just the only visible record of the New England coastal plain, but also the only evidence of the pioneer ice lobe. The shores of the Vineyard are the display cases of a unique museum, where new exhibits are brought forth and old ones discarded almost daily.

Later glaciers were to confuse and nearly wipe out definite evidence of how far this early glacier spread across New England. However, it should be noted that the Cape Cod-Nantucket area probably lay well beyond the reach of even the most far-flung of the first glacial lobes. Even the most lively of outwash streams must have been dissipated and blotted up by sands before it could reach this part of the coastal plain. Consequently, the easterly section of the pioneer ice apparently left neither tracks nor residue on Cape Cod and Nantucket, but stopped at a frontier which lay well back on the low plains, miles to the northwest and west. And these outer lands still waited to make their appearance as distinct features on the face of the earth.

CHAPTER III

Intermission

I have sent them a shell taken out of my well thirty-nine feet below the face of the earth; and I have taken many sorts of shells out of wells near forty feet down. And one time when the old men were digging a well at the stage called Siasconset, it is said, they found a whale's bone near thirty feet below the face of the earth, which things are past our accounting for. . . .

ZACCHEUS MACY
Nantucket surgeon, 1792

A Quiet Time

After the Jameco glaciation had ended, there was a long intermission before the ice was to press down upon the land again. During this quiet time the slow processes of marine deposition contributed to the structure of the Cape and Islands an impressive layer, rivaling in thickness any of the more dramatically deposited glacial segments. The Vineyard's Nashaquitsa Cliffs rise on the massive blue clay left during this period. So do Cape Cod's rugged Clay Pounds at North Truro. Nantucket's Sankaty Light is supported by a foundation of the sand left by this interstadial sea.

Altogether, these deposits tell of a pronounced and long warming of the climate which melted the ice back, well away from offshore southern New England, for several thousand years. This was probably the longest interstadial period of the Wisconsin Stage. Copious melting occurred, not only in the New England region, but along the entire periphery of the ice cap. And the seas filled. A great rising tide of waters slowly advanced over the lands from which the tide of ice had ebbed. Possibly the waters covered all of Massachusetts east of Framingham.

Frail and insignificant in the wide expanse of water, there rose a few tiny islands more than fifty miles east of the shore. They were those areas of western Marthas Vineyard, formerly under water, which the vanished Jameco ice had pushed more than one hundred feet above the level of the sea during the glaciation; never has the sea risen high enough to cover these areas again.

The Making of a Clay Deposit

As the ice melted, it left a blanket of till over the mainland. Throughout this till, with such ordinary till materials as boulders, sand and clay, there are patches of peat, lifeless and formless remains of forests. The doomed trees had spread to this area at the beginning of Jameco time as the coming ice had chilled the land, and had succumbed when ice finally passed like a juggernaut over them. Powdered through the vegetable mold and woody fibers of the peat are pollens of trees and spores of lower forms of plant life. By these pollens and spores, some of the members of the forest community can be identified. As we might expect, we find that a northern plant assortment—pines, spruces and hemlocks—had dominated these early forests, heralding by their presence the oncoming glaciation.

Then, after the ice had come and gone, swallowing the forests in its passage, the warming climate produced a southerly assortment of plants on the land surface. In vivid contrast to the trees of the peat, there now came sycamores, chestnuts, magnolias, sour gums, cedars and poplars, threading their roots through the till and the remains of their buried predecessors. Such southern species were to survive only as long as the warmth of the interstadial period lasted, and were themselves to fall with the next ice advance.

Meanwhile, streams developed channels in the tree-shaded landscape. They carved down through the till and carried large quantities of it to sea. The wave-beaten shores of the mainland also yielded their loose glacial debris. As the sea ate into the land's soft margins, it sorted the materials of the drift. Large glacial boulders, too heavy for the water to move far, remained near shore. Most of the friable peat quickly became dust. The rest of the debris was caught up in the turbulent eddies and currents of the shore zone. Drawn into deeper water, it was segregated, by varying weights, into zones of gravel, sand, silt and clay. The tiny size of pollen grains linked their fate to that of the equally small grains of clay, and together these drifted to quiet water where they settled down. A thick bed of clay formed.

The still waters which received the clay included those which submerged all but the higher hills of Marthas Vineyard and, perhaps, parts of Cape Cod. Probably this region of clay deposition stretched westward parallel to New England's southern shore, across Block Island, and all the way to the eastern tip of Long Island. Gardiners Island, within the fishtail eastern end of Long Island, has given this clay its name, Gardiners Clay, because the exposures of it there are especially thick and clear. On the Vineyard it is known simply as blue clay.

Today, since seawaters again cover much of this area, patches of the clayey interstadial sea floor are preserved to view only on these widely separated areas, from Long Island, across Marthas Vineyard, to Cape Cod.

Blue Clay on Marthas Vineyard

To see the Gardiners Clay on the Vineyard, let us start at the western end of the Nashaquitsa Cliffs and stroll eastward along the beach. At first we see clay emerge from below sea level, lying on glacial gravel. Eastward, in the higher parts of the cliffs, the same clay soars one hundred feet above the beach; in places it is the only thing clearly visible because it slumps down and covers lower parts of the cliff as well (*Figure 7*). Farther on, at the end of the cliff face, it dwindles out altogether.

Nearly everywhere the Gardiners Clay is bluish-gray, probably from disseminated organic matter. Concentrations of tiny quartz grains in some places make it white; carbonaceous powder from lignite sometimes colors it black; but these are only local variations.

If we scoop up some of this blue clay, baked dry by the sun, and squeeze it in our hands, it will crumble to powder. If we hold a lump of it in the sea, however, soon it will easily pass for children's plastic modeling clay, for when wet it can be squeezed, shaped and molded.

Close examination of clay exposures which are freshly uncovered by cliff erosion in their lower parts shows us that the Gardiners Clay is thinly layered in places. This is due to an alternation of individual clay bands about half an inch apart, etched into relief

Figure 7. *A bulwark of Gardiners Clay, Nashaquitsa Cliffs, Marthas Vineyard. This clay was deposited in quiet seawaters during an interval of ice recession.*
U. S. GEOLOGICAL SURVEY

on a tiny scale across the cliff face. Some authorities feel that such layering took place under the downward and forward pressure of ice which crossed the clay after it was deposited. Or possibly it is a result of different degrees of glacial melting during different seasons; heavy summer melting would lead to thicker layers of debris than slow winter melting. The thicker layers of coarser material would withstand subsequent erosion best and would become etched into relief.

The best Gardiners Clay samples are at Gay Head and Menemsha. But it appears elsewhere as well. From the northern end of the Gay Head cliffs there are patches of it continuing northeastward all along the shore of Vineyard Sound to Norton's Point, but not beyond. It is easily spotted west of Prospect Hill in the bluffs, and turns up again near the mouth of Roaring Brook. Here it was used during the last century for china clay and brickmaking. It is said that a few remnants, such as bits of smokestacks and waterwheels, remain of the old factories, and can be spotted if you search diligently along the brook. Lack of fuel forced them to stop working in 1870, after years of providing mainland cities with thousands of bricks annually.

Strangely, Gay Head's cliffs have no Gardiners Clay in their central part, even though it shows up so well just east of Gay Head in the Nashaquitsa region. And at the northwest end of Gay Head too, there is Gardiners Clay; it lies partway up the cliff on earlier preglacial and Jameco glacial deposits. From this place the clay stretches eastward along the island's shore, the beds sloping away from their Gay Head exposure. Thus the central part of Gay Head stands as a massive island, with Gardiners Clay lapping only partway up its western and eastern sides.

Why is the clay lacking in the central part? It has been postulated that Gay Head was an island at the time the clay was deposited. The relatively quiet waters in which the clay grains drifted down lapped partway up the higher parts of the Vineyard's preglacial and early glacial foundation. The very presence of these higher parts as islands or part-time shoals may have helped shield and temper the seas from storm winds which otherwise could sweep across open water and prevent the clay from settling. Wherever water covered the submerged parts of earlier deposits, it blanketed them with clay. Much later, when the waters were to withdraw again, hills such as Gay Head, which probably had risen as islands from the sea, would appear as islands still, rising above a sea of clay.

Blue Clay on Cape Cod

In places on Cape Cod, Gardiners Clay makes a dramatic appearance. We go to Cape Cod Light, North Truro, descend to the beach and walk north. Soon we reach a place where the central third layer of the cliffs is solid blue-gray clay, capped by tan glacial drift and some twenty feet of sand and silt. The clay here is twenty-five feet thick. From it the region takes its name, Clay Pounds.[4]

The clay is hard when dry, easily molded when wet; and it often develops small pellets on its surfaces like the top of a crumb cake. Beneath the clay is tarnished gravel, usually partly hidden by loose debris.[5]

At the turn of the century, geologist Myron Fuller published a report on the economic possibilities of Cape Cod's clay, of which the Gardiners Clay at Cape Cod Light makes up the largest exposed mass. Fuller suggested that "it is not impossible . . . that at some future time, when more accessible deposits elsewhere are exhausted, the clays [near Cape Cod Light] may be worked with profit if connection by tramway or otherwise is made with the inner side of the cape, from which shipments could be made by vessel." Obviously the value of shore front real estate was not what it is today.

Apparently, after the sea had drawn back to yield to the ice which followed the Gardiners interstadial interval, this thick layer of clay was humped up by ice pressure—"erected in the midst of sand hills, by the God of nature," as Levi Whitman suggested in 1795, "on purpose for the foundation of a lighthouse."[6]

Lack of Blue Clay on Nantucket

On Nantucket, no Gardiners Clay has been identified. As we shall see, conditions were different in that region, and Nantucket received a unique layer of its own, quite different from the blue clay.

[4] It has been said since very early days that the name comes from the way the sea pounds against the clay bluffs during storms, but Thoreau thought it more likely referred to the many clay-floored ponds scattered on the surface of the uplands near the cliffs—"clay ponds" having become distorted to "clay pounds."

[5] A controversy exists concerning this clay. Some geologists hold the view that it is not marine Gardiners Clay but belongs to more recent glacial outwash. Evidence available at the present time indicates to the writer that the Clay Pounds clay is identical in age and origin with the Vineyard's Gardiners Clay. The interested reader is referred to references by E. Hyyppä, 1955; R. W. Sayles and A. Knox, 1943; and J. B. Woodworth and E. Wigglesworth, 1934, in the bibliography.

[6] Three years later, in 1798, the first Cape Cod Light was erected on the Clay Pounds.

Fossils in the Blue Clay

Remains of seashells are sparse and unspectacular in the New England Gardiners Clay. In its presumed continuation on Long Island, however, there are many shells of animals similar to those which today inhabit New England's offshore waters. Some shelled creatures range far and wide. Others are sensitive to variations in water temperatures. They spend their lives caged behind invisible bars of small temperature changes. The shells of such animals in the Long Island Gardiners Clay indicate that the temperature of those interstadial waters and therefore, probably, of the air above them, was similar to that of the region today. To so have lost its grip on the climate, the Jameco ice must have withdrawn far to the north during the clay interval.

But the Cape-Island blue clay is not wholly barren of fossils. If you powder it up and examine the grains carefully with a good magnifying glass, signs of life may emerge. There are many tiny siliceous sponge spicules, the hard parts of the sponge animal. These take various shapes: they may be needle-like, V-shaped, or look like two- or three-pronged anchors. Apparently sponges led a rather lonely existence on this oozy sea floor.

However, the currents in the waters above carried life, among which were diatoms. These are microscopic one-celled plants of the algae family, common to all waters today. When they died, their shells fell to the bottom to join the sponge spicules in the clay. Under a lens these would appear as tiny geometric figures, perhaps crossed by a delicate, lace-like network. Diatom shells are hard and siliceous. Chemically they are opal, although they have no outward resemblance to precious opal.

When diatoms from the northern blue clays are compared with those from the same clay on Long Island, there are some variations. Among the northern diatoms are some species which lived in relatively cold waters. This may reflect the fact that the Cape-Island region was nearly a hundred miles nearer the retreating ice front than Long Island.

Still another type of fossil appears in Cape-Vineyard Gardiners Clay—the pollen from the peat which dates back to the coniferous forests which had formed the advance guard of the first Jameco ice.

So great is nature's regenerative drive that these vital kernels of plant propagation were shaped and armored for a seemingly indefinite survival—and have survived for some fifteen thousand years.

Nantucket's Shelly Sand

Nantucket Island has a distinctive bed of sand which takes the place of its neighbors' Gardiners Clay. Extremely variable in color, this sand is streaked with motley patterns of white, brown, red, yellow and gray. It shows up best today, somewhat tilted and distorted, a few hundred yards south of the lighthouse in the bluffs which line the shore from Sankaty Head to Squam Head. Its name is the Sankaty Sand.

Throughout the nineteenth century wave erosion kept this sand freshly exposed in the cliff, and it readily could be examined. During recent years, however, a wide shelf of beach has developed in front of the cliffs, shielding them from wave attack. Thus protected, beach grasses root themselves on the talus which drapes the lower part of the bluffs, and much of the Sankaty Sand is hidden beneath a green carpet.

The sand consists mainly of mineral fragments such as quartz, with clay and iron oxide, like crumbling mortar, between them. Toward the base of the visible section of the cliffs there are bits of green glauconite and fragments of garnets. Higher up, bluish clay appears, and occasional rock pebbles. These and the minerals are clearly of sorts common in rock formations of southeastern New England and its continental shelf.

The oyster bed comes next. This is a nine-inch layer of beach-like sand replete with fossil shells. It links us to life in the Pleistocene sea, for these shells remain nestled in the sand much as if they still lived. Oysters are very common, their shells closed as if in life, and with barnacle shells still clinging to them; there are quahog shells in which even the horny valve ligament remains and holds the shells closed; the shells of steamer clams, closed, still sit upright; and there are tiny snail shells which show no sign at all of wear by waves or currents.

Above this layer is the so-called Serpula bed, a thicker, two-foot bed with remains of the shelled sea worm, Serpula, and the riddled shells of molluscs on which these worms fed. In some places this layer is so full of shells that there is room for scarcely any mineral grains.

The remaining beds above these are altogether about two feet thick. They contain still more shells and fragments. These, however, differ from the lower insofar as the species are definitely more northern in habitat. There is a hodgepodge of whelks, barnacles and mussels toward the top.

The Sankaty Sand apparently is contemporaneous with the Gardiners Clay. As we have seen, clay grains slowly dusted the quiet interisland waters just off the coast. Meanwhile, some disposition had to be made of the coarser sand which likewise had been washed from the mainland's till. It piled up on adjacent parts of the sea floor. Barrier beaches built by the sea may have created quiet lagoons between themselves and the gravelly surface of ancestral Nantucket during interstadial times. In these, clams, oysters and snails found refuge and throve, as the same creatures thrive today in similar surroundings.

Of course, there must have been some sand which first fell to the bottom in an offshore zone just west of the clay belt and there became inhabited by sea floor communities. Prodded up later by active currents, some of this sand soon must have been swept away to the east, with many shells broken in the process. Past the clay zone this debris drifted, to an arc-shaped belt swinging northeast from Nantucket as far as Boston.

At that time there were at most only tiny islands in this eastern region. The margins of Nantucket, parts of Cape Cod and the fishing banks were covered by shallow interstadial waters. Gradually the sheet of sand, the Sankaty Sand, covered the sea floor and grew in thickness to more than one hundred and sixty feet.

Obviously life found this sand-floored area vastly more hospitable than the clay region to its west. Through the loose, unresisting sand, plants could send their roots and grow, and wherever plant life pioneers, animals are likely to follow. The sea creatures throve. They reproduced their kinds and died in such fecund abundance that today patches of sand are filled with their whole and broken shells.

Generally, of course, it was the most stable of environments to any one of these creatures. But to the community as a whole, it was the most transient. Again the glacial tide was waxing. The warning signs were everywhere. An ominous lowering of water temperatures began to make itself felt. Life zones again began their slow southward shift. Some of the creatures, such as clams, apparently remained, unaffected by the temperature drop. But all around them their warmth-loving neighbors moved out and Arctic forms moved in. Creatures of the sorts which now live in frigid seas began to make their homes on these Sankaty sediments. It would be many years before the resurgent ice actually would reach this area, but as we observe the difference in fossils in the upper part of the Sankaty Sand from those in the lower, we are warned, in retrospect, of its approach.

For centuries the loose, shell-strewn sand lay beneath the sea. Tidal waters washed it. Tidal currents rippled its surface, as near-shore sand today becomes ridged and rippled by the water. Thousands of years have passed since then, and the sand has been lifted out of the sea, but these ripple marks remain in some places and sometimes still may be seen in the lower exposed part of the deposit.

Sankaty Head has the only Sankaty Sand visible on Nantucket. However, under most of the island there stretches an irregular layer of the same material. As early as 1792, some was found by Zaccheus Macy. Diggings for a forty-foot well on his property near Nantucket Town brought up pieces of shell which Macy reported to the Massachusetts Historical Society (*see page 47*). It is not unlikely that the whalebone to which he refers also came from the Sankaty Sand. Other islanders were discovering the same thing. Walter Folger reported in 1792: "There have been many times found at the bottom of wells, at the depth of forty and fifty feet, and after digging through several strata of earth, such as clay, &c., shells of the same kind as are now found on the shores of the island. . . ."

The Jacob Sand

To go back in time to the temperature drop which divided the Sankaty Sand into two distinct faunal zones: things were happening in the region of Gardiners Clay deposition as well. Capping the clay here and there we find large patches of coarser sand with cold-water type shell fossils. It is not hard to reconstruct events. With the approach of the ice, the quiet water conditions necessary for the accumulation of the clay came to an end. Doubtless the sea became chilled and shallower. Winter blasts churned its waters, and possibly its currents shifted. A fresh influx of fine sand, probably washed to the sea from the front edge of the coming ice, drifted into the shallow waters. Much of the clay which formerly would have settled to the floor in the clay area now was swept away by the turbulent waters, and in its place coarse sands fell. These soon covered the barren clays that already lay on the bottom.

This sand has been named the Jacob Sand. At one time it must have covered all of the Gardiners Clay, but only tattered remnants have survived the glaciers. In some places it is fifty feet thick.

The Jacob Sand grains are larger than those of the dusty rock flour which makes up the Gardiners Clay. However, the two formations look very much alike in places, for the grains of Jacob Sand are much smaller than those of the Sankaty Sand on Nantucket, small enough to bind together and make the Jacob Sand somewhat plastic, like the Gardiners Clay, when wet. And large parts of both formations are light gray. In places the Jacob Sand shifts to yellow-orange and looks like fine beach sand. In fact, yellow sandy layers commonly alternate with gray clay layers in the Jacob Sand.

The iron which colors these sandy layers yellow is continuously leached from them and carried downward by rainwater. When the water reaches the massive, impermeable Gardiners Clay beneath, it stops and deposits its iron. In some cliff outcrops of these beds, a rusty crust of iron often appears at the contact line between the Gardiners Clay and the Jacob Sand. Nashaquitsa has the best exposure of the Jacob Sand, riding on the Gardiners Clay above the iron band. Here the looser layers of yellow sand are more affected by wind than the compact clayey layers of the Jacob, so that the latter have become etched into relief like architectural flutings.

The fossils of the Jacob Sand tell us that as it blanketed the clay of the sea floor, marine life spread to this region and left shells in the sand. Farther south, Long Island's Jacob Sand bed has a great many such fossils, although we have to search carefully in the Vineyard's Jacob Sand to find shells. What have been found on the Vineyard and on Long Island are cold-water varieties, types of snails, scallops and clams similar to those which now live in the Gulf of Maine.

Thinly veneering the Jacob Sand we can find a very fine windblown grit, like the fine grit which caps some sea cliffs today. Here is evidence that before the ice reached it, the sea floor became land again, its waters withdrawing to become snow and to nourish the swelling ice sheets. Vast low-lying plains of sand lay exposed off the southeastern New England shore. The violent winds which whipped about the front of the coming ice swept up the loose surface grains into the air again and again, shooting them against one another and dropping them pitted and pulverized.

Then came the long burial of these plains, first beneath the ice, later beneath its drift and the waters of the sea. So are they buried still, except where waves have exposed them in cliffs—slices of an exhumed sea bottom, dried and alien in today's bright sun.

Looking backward briefly now at the long Gardiners interstadial interval, we see the melting back of the Jameco glaciers and in their place a rising sea. With the withdrawal of the ice across southern New England, we see warm-climate forests replacing the northern species that had edged the ice on its early advance. On the mainland, eroding rivers course across the forested till landscape, carrying their loads of clay, sand and vegetable matter to the quiet sea which already covers most of the region from Cape Cod to Long Island. In this sea the sediments pile up: fine clay—Gardiners Clay—in the Cape Cod-Marthas Vineyard region, and coarser sand—Sankaty Sand—on Nantucket. On these sediment-covered sea floors, particularly on the more hospitable sand, sea life takes hold. Through the centuries the changing communities of life mirror the changing climate. The shift of species from warm to cold types tells us that glaciers are readvancing. At the same time, in the Vineyard region, we no longer find clay drifting down in quiet seas. Instead we feel cold winds and turbulent weather changes: and the soft Gardiners Clay becomes blanketed by gritty Jacob Sand, which drifts into the region from the north—from the front of the oncoming, overpowering ice.

CHAPTER IV

Red Ground and Sky-Blue Paint

The conditions that prevailed during the Montauk stage of the Manhasset were those that are associated with a vigorous continental ice sheet. The area that now forms Marthas Vineyard was covered by a thick glacier, which extended at least to the southern limit of the island and probably beyond. . . . The region must have remained buried under the ice for a long time, for the till laid down is thick and there is nowhere any evidence that it accumulated rapidly. Toward the end of this stage there naturally must have been a stage when the ice was stagnant and melting. . . . After this great ice sheet had disappeared the region must have presented a picture that surpassed in desolation and barrenness anything now in existence.

EDWARD WIGGLESWORTH
Geologist, 1934

Nature of the Manhasset Ice

The pioneering Jameco glaciers had been but a preview of the great Manhasset glaciation. It was as if the ice had come south to make a preliminary reconnaissance of the region before strengthening itself during the long Gardiners interstadial interval for its major invasion of southeastern New England.

The glaciers of this Manhasset invasion strewed abundant evidence of their presence on Cape Cod and on both Islands. Beneath this mighty composite ice sheet and along its swollen meltwater streams, Cape Cod and the Islands received much of their present bulk. Tracing the Manhasset deposits, we can see that this ice was the sculptor which preliminarily blocked out the forms of these lands, roughly anticipating their present shapes.

The outlying portions of this powerful ice of the Manhasset substage sent forth three glacial lobes from the southeastern New England mainland. These were molded by the lay of the land and the climatic uncertainties born of the sea. The advance margin of the glacier doubtless forked around New Hampshire's White Mountains, emerging as two lobes. Another barrier may have been Mount Katahdin in Maine, an obstruction reinforced by lower Mount Desert to its south.

The exact contours of the area which this second glacier covered are uncertain, for its deposits were scattered and partly obliterated by the third and final ice invasion (Chapters V, VI). By and large, however, evidence indicates that the position of the Manhasset ice conformed somewhat to that of the final glaciation which followed, shown in *Figure 11*.

On their way south, the Manhasset lobes moved in somewhat different directions and at different speeds. Southward from Maine came the most seaward of the three, a powerful ice mass. It filled in and spilled far over one of the now-drowned coastal-plain valleys, which swings northward parallel to the coastline. The landward margin of this lobe lay slightly to the east of Cape Cod's forearm. This ice lobe may have reached as far south as the latitude of Nantucket Island. It is named the South Channel Lobe, taking its name from the deep, thirty-mile-wide South Channel between the submerged Nantucket Shoals and Georges Bank. However, it did not reach South Channel, probably, until during a glaciation which occurred after the Manhasset and during which the lobe readvanced.

The South Channel Lobe during the Manhasset, moving entirely across what is now sea floor, gave to this submarine area much of its present shape. It deepened the valleys. Piling the spoils on the ridges, it raised them far above the levels of the simple coastal-plain hills they once had been.

The westerly neighbor of the South Channel Lobe was the middle of the three ice masses, the Cape Cod Bay Lobe. It moved south-southeast. It probably left the mainland by cutting across the bedrock ledge which juts out from Massachusetts as Cape Ann. It enlarged a preglacial valley now filled by the waters of the present Cape Cod Bay. Like the South Channel Lobe, the Cape Cod Bay Lobe probably reached only as far south as the latitude of Nantucket.

Still farther west, nearest the mainland, crept the third of the lobes, the Buzzards Bay Lobe. This traveled more nearly eastward, across southeastern Massachusetts and possibly Rhode Island. It piled some of its till on today's Marthas Vineyard and reached an unknown distance beyond the Vineyard's present shores, fanning out radially as it moved to as far south as Long Island.

This entire Manhasset glaciation, like the Jameco preceding it, was a substage of the Wisconsin glacial Stage. The gravels and tills remaining from it are known collectively as the Manhasset Formation.

Manhasset Ice on Marthas Vineyard

The westernmost, Buzzards Bay Lobe, was probably the first of the three to reach the latitude of Marthas Vineyard. The coastal plain over which it passed had changed from a quiet, clay-covered sea floor to a chain of chilly, wind-whipped island sand plains, rising from the water only a few feet and strung out eastward at least as far as Nantucket Island. Torrential summer freshets and sluggish winter streams flowed from the ice front and laced across the flat plains, carrying sand and gravel. So laden with debris was the ice that the outwash materials thickened in places to more than a hundred feet before the ice itself finally overrode them.

This outwash is Herod Gravel. We can see it at Nashaquitsa, a coarse pebbly material which lies above the finer-grained Jacob Sand capping the Gardiners Clay. Here the Herod Gravel is one hundred feet thick, although it is greatly folded and wrinkled. Lesser amounts of it show up, likewise above the Jacob Sand, at Gay Head, with some too at Stonewall Beach and in the cliffs along the island's northwest coast (*Figure 8*). Its extent indicates that the outwash plain once must have covered at least the entire western part of the island, and possibly the central and northeastern parts as well.

Following its meltwaters, the ice itself arrived. It distorted by its movements the pre-Gardiners Till and the Gardiners Clay, and it piled its own load upon all pre-Manhasset deposits.

If you go to some morainal part of the Vineyard—Chilmark, for instance—you can dig down through time. First your spade sinks through the dark topsoil; strikes perhaps a rock in the bouldery moraine left by post-Manhasset ice; grates through the thick gravelly outwash beneath it; and finally stops, nosed against an unyielding material that farmers call hardpan. It is the Montauk Till, the name

Figure 8. *Till on outwash gravel at Marthas Vineyard. Pebbly Montauk Till lies on the layered Herod Gravel at Stonewall Beach.* U. S. GEOLOGICAL SURVEY

given to the till left by the Buzzards Bay Lobe during its Manhasset visit to the island.

When first deposited, the Montauk Till was a bouldery clay not greatly different from the upper till at Chilmark. However, long years of burial beneath the Manhasset ice mass followed—and later, more long years beneath a resurgent ice sheet which piled the uppermost till upon it. The icy weight squeezed the clay grains of the till into a tightly fitted interlocking pattern. Some of the grains even may have recrystallized and glued themselves together. The entire process was something like the way a glacier's lower layers become ice. Finally, chemical solutions brought by trickling groundwater have deposited iron oxide, which has acted as a binder and completed the natural concrete. In fact, those who sink wells to a layer beneath the Montauk Till get considerable amounts of iron in their water, dissolved from the iron-rich till and carried by seepage to the sediments beneath.

At Squibnocket Point, chocolate-brown and gray Montauk Till streaks across the central part of the bluffs. Coursing waters have carved strange cliff fantasies into it. Pointed spurs, steep gullies and eroded peaks provide all the appurtenances of a badlands (*Figure* 9). Dotted through it all like raisins in a pound cake are some large boulders up to six feet across and many small ones, all torn from the mainland. Undermined by water, they fall from the cliff and spread patches of cobblestones over the beach.

Figure 9. *"Badlands" erosion in the Montauk Till, Squibnocket Cliffs, Marthas Vineyard.* U. S. GEOLOGICAL SURVEY

Less dramatic pockets of Montauk Till sometimes show up at Stonewall Beach (*Figure 8*), at Gay Head, and here and there along the northwest shore. At Tisbury's Cedar Tree Neck, part of the cliff shows nothing but this till. Unlike Squibnocket's strange badlands, the land here terminates in plain vertical cliffs, for rainfall which gathers at the crest of this bluff flows away from it to a small pond behind, rather than picturesquely down its face.

The Montauk Till is about forty feet thick. Bluish-gray in places and doughy with clay, it is hard to distinguish from the Gardiners Clay farther down, or even the Jacob Sand between the two. Naturally it contains much of both these formations, for the ice eroded them as it traveled. Here and there iron has stained some of the upper part of the Montauk brown; in other places, carbon makes it gray and black.

The visible occurrences of till from place to place across the Vineyard indicate that a layer of it must spread over most of the island. As far east as Edgartown and Chappaquiddick you may hear tell of "red ground" which, as impure clay, has gone into some of the old roads. This is Montauk Till, locally red from iron oxide.

As the ice began to edge away from the Vineyard, it gradually released the surface, bleak and boulder-strewn, to the air. Meltwater streams, drawing back with the ice, found their way across it. This resulting outwash gravel is called the Hempstead Gravel.

It is identical in appearance and composition with the Herod Outwash Gravel (previously mentioned) which had preceded the ice advance. The Montauk Till thus originally was sandwiched between the Herod and the Hempstead. Resurgent ice, however, has stripped away most of the Hempstead so that it is rare on the Vineyard today. Forty feet of it appear at Cape Higgon, where it extends laterally some five hundred feet above the stony Montauk Till of the bluffs. It is impossible to tell how thick it was originally. Probably the gravel at the base of West Chop and East Chop is Hempstead Gravel; and some appears along the north shore of Chappaquiddick.

Meltwaters on Nantucket

Nantucket Island lay so far south and east during the Manhasset glaciation that it again missed out on a coating of till. We should expect to find such till just above the Sankaty Sand. But if we go to Sankaty Head and examine the cliff, we find no till. Instead there are twelve feet of coarse gravel. The individual pebbles are not the white and tan quartz which we are used to finding in gravels, such as the Vineyard's Aquinnah Conglomerate. Instead they are multicolored irregular rock fragments. These rocks are distinctive enough to trace back to their places of origin on the mainland. Most are from southern Massachusetts and northern Rhode Island.

This obviously is outwash gravel. It makes up the only drift left by the Manhasset ice on Nantucket's outlying plains of sand. Although the ice itself seems never to have reached Nantucket Island, meltwater streams did, and they left their gravel deposits.

Manhasset Ice on Cape Cod

Let us go back in time to when the front of the Buzzards Bay Lobe was lingering somewhere beyond Marthas Vineyard, and consider the rest of that glacial trilogy, the Cape Cod Bay Lobe and the South Channel Lobe. For they also were active and powerful, although not as tenacious in their grip on the land as the Buzzards Bay Lobe. While the first powerful thrust of the Buzzards Bay Lobe piled its debris over the Vineyard, the other ice in the less stable maritime environment to the east apparently pulsed back and forth three times, leaving three layers of drift. Cape Cod's present northern shorelines may be the show windows for this debris.

We stand facing the sea on Eastham's Coast Guard Beach, below the old Coast Guard station and at the low part of the bluffs in front of the public parking area. In the bank to our right, short horizontal layers of earthy material jut out like discontinuous ladder steps beyond cascades of white and brown sand. One such layer occasionally has been visible, directly at the base of the bluffs, for easy examination. Its flat top extends level from the bank, capping some three feet of chocolate-colored, sticky-looking material. We can see even from a distance that this layer is full of pebbles spattered through in all positions as if they had been blasted from a shotgun. It is typical till. The matrix supporting the pebbles is clayey, with holes here and there from which other pebbles have dribbled down to the beach. Other layers much like it are exposed higher in the banks (*Figure 10*).

The step-like pattern of the tills is due to their high concentration of clay, including much Gardiners Clay. Cohesive and compact, such clay withstands erosion. Between the till layers are sandy intertills. Each time the ice drew back, these were deposited in the swelling seawaters. Such intertills are loose and crumbly. Wind and water easily chisel them away from the tills, leaving the latter to stand forth alone.

Altogether there are three till layers from this glaciation. From lowest to highest they are known as Till 4, Till 3 and Till 2 (Till 1 is above them and represents a later glacial substage).

Figure 10. *Step-like layers of till at Nauset Beach. These result from a fluctuating ice sheet. Between the protruding pebbly tills are sandy intertills, laid down in seawaters between ice advances.* B. CHAMBERLAIN

We were introduced to these tills at Coast Guard Beach because this is probably the best place to get a closeup look at one. To see their large-scale pattern, however, we can go north the short distance to Nauset Light Beach. Then from the parking area beneath the light, let us walk still farther north along the beach. In the cliffs we usually can make out three parallel layers of tan, clayey, pebbly till separated by intertills.

Sometimes when storms have taken a fresh sampling, they are so well exposed north of the lighthouse that their entire structure is visible as a broad, very gentle uparching. Other times, however, the bedding plays hide-and-seek, and can be seen only where occasional portions have not been veiled by loose cliff waste.

In fresh exposures the nearly flat bedding of the sandy intertill layers is visible, paralleling the tills. Just as every till in vertical sequence represents one forward surge of the wavering ice margin, so every intertill layer of marine deposition represents a retreat.

The intertills all look very much alike. So do the tills, since their compositions are nearly identical. Unless the complete stack of them is exposed they cannot be told apart. It seems that the ice followed almost the same route in each of its three advances and picked up very nearly the same materials.

Altogether, the small-scale comings and goings of ice and sea during this Manhasset glaciation left at least thirty-five feet of sediment on the Cape. Much of this debris came from southwestern Maine, according to the makeup of the rocks and pebbles in it. Thus the ice flowed nearly due south each time it advanced. Sometimes the tills show a strong family resemblance to one of their direct ancestors, the Gardiners Clay, scraped up from the present sea floor. We can assume that at the time of these advances, miles of the present offshore area lay exposed to air. Even on the submerged parts of the continental shelf the ice was aground, for with its several thousand foot thickness it crept easily through the then-shallow water.

A fourth drift layer, Till 1, runs along the top of the banks at the Nauset cliffs. Beneath it is an intertill layer which separates it from the uppermost (Till 2) of the three Manhasset tills.

This highest intertill is significant. The lower ones, like succeeding pages in a story, reflect minor fluctuations within a single glaciation. The highest intertill introduces an entirely new chapter, for it represents an interval between two major glaciations.

The Manhasset tills and intertills are so common that they must underlie most of eastern Cape Cod at least. They do not always look exactly like those at Nauset; in some places the tills are as sandy as intertills; in other places they may hold boulders up to six feet across.

We can see them in almost any part of the bluffs along Cape Cod Bay from Dennis to Truro. Here, as on the ocean side, slumping often obscures the materials beneath, but dark horizontal steps of till always show up here and there. Indian Neck, Wellfleet, and Nobscusset Beach, Dennis, usually have good exposures. Here the fresh till looks like pebbly gray clay.

In some places Till 4, the lowest, lies below sea level. At Brewster what is probably Till 4 lies just beneath the sandy surface of the tidal flats which are bared at low tide. Parts of higher tills show up in the bluffs, especially after storms. Within the past decade several new rock groins have appeared along Brewster beaches, reaching into the water from about high-tide line. Upon completion of each, little blobs of blue-gray clay till began to ooze up through the sand beside the groin, squeezed like toothpaste by the downward weight of the rocks on the till under them.

Fossils found their way into the Manhasset ice. Examine carefully a sample of clayey till with a magnifying glass and you may see plant and animal remains similar in type and proportions to those of the Gardiners Clay. They must have gotten into the ice as it scraped up parts of the sea floor on its travels. Among these are pollens of cold-climate plants—a third pine and spruce, the rest northern types of nonevergreens. These presumably are from forests that had succumbed to the earliest, Jameco glaciation. As we have seen, they formed peat, which then was eroded from the land and mingled with the sediments on the sea floor during the Gardiners interval. There also are siliceous marine sponge spicules and diatom shells in the till, of types that may have lived during the Gardiners interval when the waters were warm and quiet.

Cape Codders from the earliest days to today have used the ever-present clay tills. Cape Indians, by mixing the clay with shell fragments and sand, made delicately thin, light pottery. The same tills went into road building during the last century; part of the Wellfleet section of the old Cape highway was built from this impure clay. About this time some of it was found to be tolerably good for brickmaking. A thriving local industry sprang up. Clay dug from West Barnstable and Town Neck, Sandwich, was mixed with sand and turned into enough bricks to meet local needs for well over a century. And the familiar sky-blue paint with which early Cape Codders liked to cover their wagons came from mixing the blue clayey tills with skimmed milk.

In the homogenization of the nation and its industries, such local endeavors died. But the tills are still used by hobbyists who collect them locally and fire them into ceramics. Although not all of them are good for this, the purer ones take firing well. Chemical changes in the kilns give the finished work a rich red-brown color.

End of the Manhasset

The Manhasset ice lobes met their end with a brief but pronounced warming of the climate. The South Channel Lobe and the Cape Cod Bay Lobe lost their identities as they dwindled far inland, perhaps all the way to the Great Lakes. Seawaters rose again. In them a new intertill came into being and blanketed Till 2. At the same time the Buzzards Bay Lobe must have retreated too, although probably not as far as the other two.

As we have seen, the three great Manhasset ice lobes contributed much to the substance of Marthas Vineyard and Cape Cod, and a little to Nantucket. At first, from the western Buzzards Bay Lobe, we saw at least a hundred feet of Herod Outwash Gravel pour across western Marthas Vineyard. Then arrived the ice itself, and it left a thick layer of Montauk Till across the island. At the same time, to the east, we watched the central Cape Cod Bay Lobe and the easternmost South Channel Lobe thrice swelling and waning across the site of Cape Cod. With each advance they left a layer of till; with each retreat they allowed the sea to occupy the region and deposit a nonglacial layer of debris, or intertill. Meanwhile, meltwater streams were pouring southeastward from the ice lobes; they channeled across the coastal plain and piled gravel on the site of Nantucket. Finally we watched the great Manhasset glaciation draw to a close; we saw the ice slowly waste backward and expose the Cape and Vineyard, while its meltwaters threaded across the surface of the till it had left behind. From these meltwaters the Hempstead Outwash Gravel piled up on the Vineyard, burying much of the thick clayey till; and on Cape Cod a final intertill capped the triple-layered sequence of tills and intertills.

Southeastern New England was not to be released for long. As we shall see, ice was to return, thick and powerful. It was, as the final ice to cover the region, the sculptor of the landscape—the scenery-maker. Its effects would be the most dramatic of all. It would wrinkle the foundations of the land, crush its surface and spill out over it its sand-filled meltwaters. It would heap hills and ridges on Cape Cod and the Islands, gouge out pits and valleys and dump over everything a disordered array of boulders. Then, after it had drawn back for the last time, the kindlier ministrations of sun and warmth would bring life to the land, and dress its wounds.

CHAPTER V

The Scenery-Maker

I – THE ISLANDS

. . . in the hills of . . . Nantucket [and] Marthas Vineyard, . . . we have one of the most remarkable true terminal moraines anywhere to be found in the world. Throughout their whole extent, these terminal accumulations form a marked feature in the landscape, rising for a considerable portion of the distance from 150 to 300 feet above the general level of the country, and being dotted over with huge boulders transplanted a greater or less distance from the north.

G. FREDERICK WRIGHT
Geologist, 1890

For the last time, ice overwhelmed New England. At the farthest limit of this advance, it formed most of the Islands. Their hills are parts of its terminal moraine; their plains are outwash plains. And as the same ice finally melted back, it stopped long enough to form most of Cape Cod (*Figure 11*). As we shall see in the next chapter, the Upper Cape's hills are part of the recessional moraine; and its plain is outwash plain.

On this final advance, the glacier once more fell into the lobate pattern which had characterized its previous advance. Outlining the entire ice mass along its southern edge, there developed a long ridge of terminal moraine. Broad festoons mark the trend of this moraine, for its parallels the scalloped, lobate margin of the ice sheet (*Figure 12*).

The Terminal Moraine

On Long Island, New York, the terminal moraine is called the Ronkonkoma Moraine. It runs along the backbone of the island to

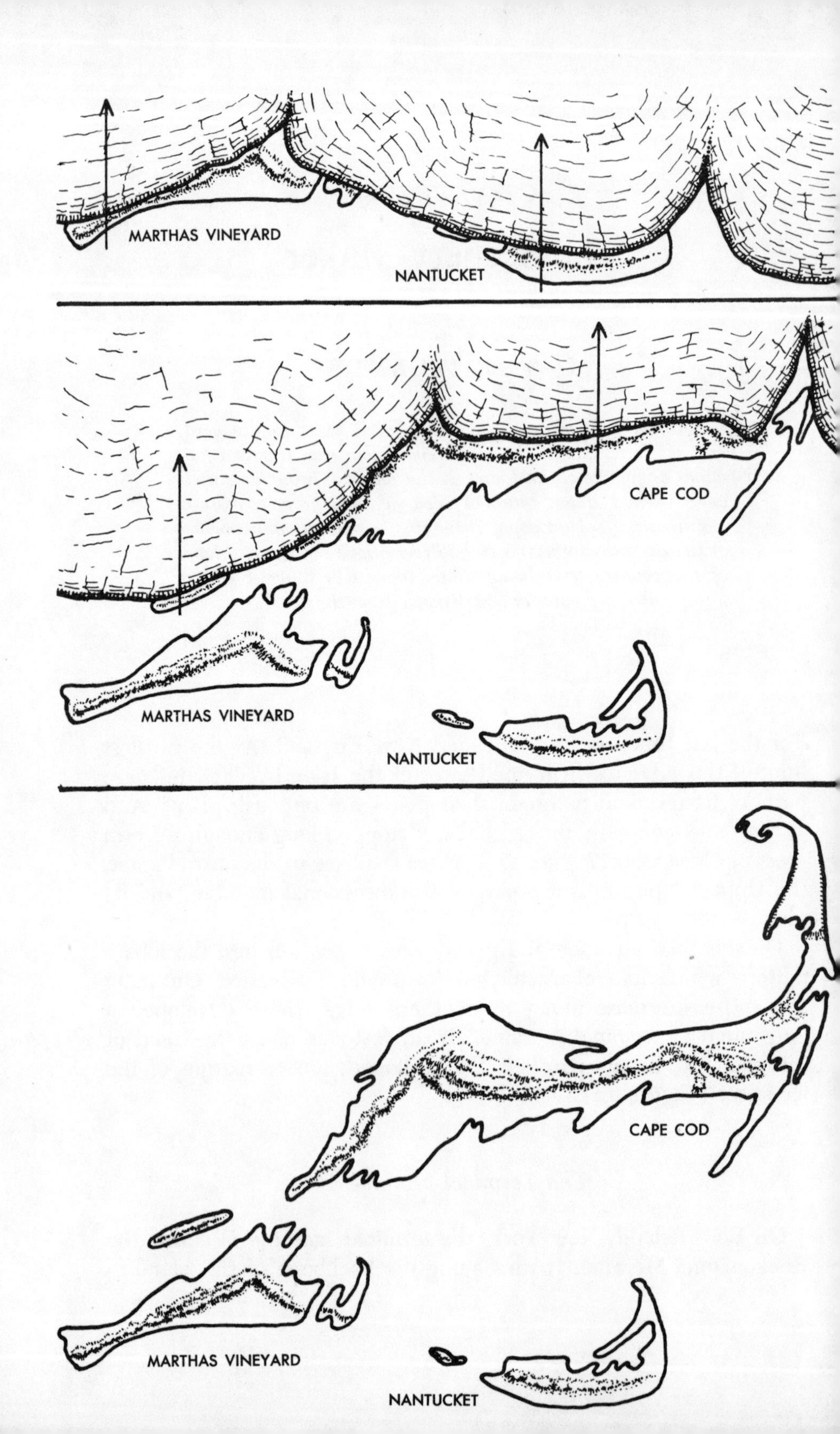
MARTHAS VINEYARD
NANTUCKET
CAPE COD
MARTHAS VINEYARD
NANTUCKET
CAPE COD
MARTHAS VINEYARD
NANTUCKET

meet the sea in Montauk's boulder-studded cliff. The Ronkonkoma Moraine is the product of an ice lobe which lay farther west than the Buzzards Bay Lobe.

Eastward from Montauk the terminal moraine loops like a playful porpoise beneath and above the water in a gracefully curved path. We need submarine soundings to detect it ridging the sea floor before it emerges as the hills of Block Island. Then it dips again below the water east of Block Island. It does not reappear until it makes up the hills of Marthas Vineyard, then those of Nantucket. Finally it dips under the sea just far enough to create the Nantucket Shoals and Georges Bank.

In the Cape-Island region, all of the most recent ice deposits come from three lobes which had the same general pattern as those of the Manhasset advance—a Buzzards Bay Lobe landward; a South Channel Lobe seaward; and a Cape Cod Bay Lobe between the two. It was from the Cape Cod Bay and Buzzards Bay Lobes that the Vineyard received its final glacial refinements.

Ice Deposits on Marthas Vineyard

Marthas Vineyard is a triangle (*Figure 13*). It owes its shape to its position within the scalloped border of these two lobes. The ice edge, so to speak, was the mold, and the Vineyard was the cast inside it, remaining long after the mold had vanished. The Vineyard's hill belt is the terminal moraine. It makes up the two northern legs of the triangle.

The Buzzards Bay Lobe contributed the northwest of these legs, the hills which run southwest from the tip of the island near Vineyard Haven to Gay Head and across to the small island of No Mans Land. The Cape Cod Bay Lobe contributed the hills parallel to the northeastern leg, running southeast from the head of Lake

SEE OPPOSITE PAGE.
Figure 11. *The glaciers on Cape Cod and the Islands.*
TOP: *Building of the terminal moraine across the Islands.*
CENTER: *Building of the recessional moraine across Cape Cod.*
BOTTOM: *After the ice had left and the sea had been at work along the shores.*

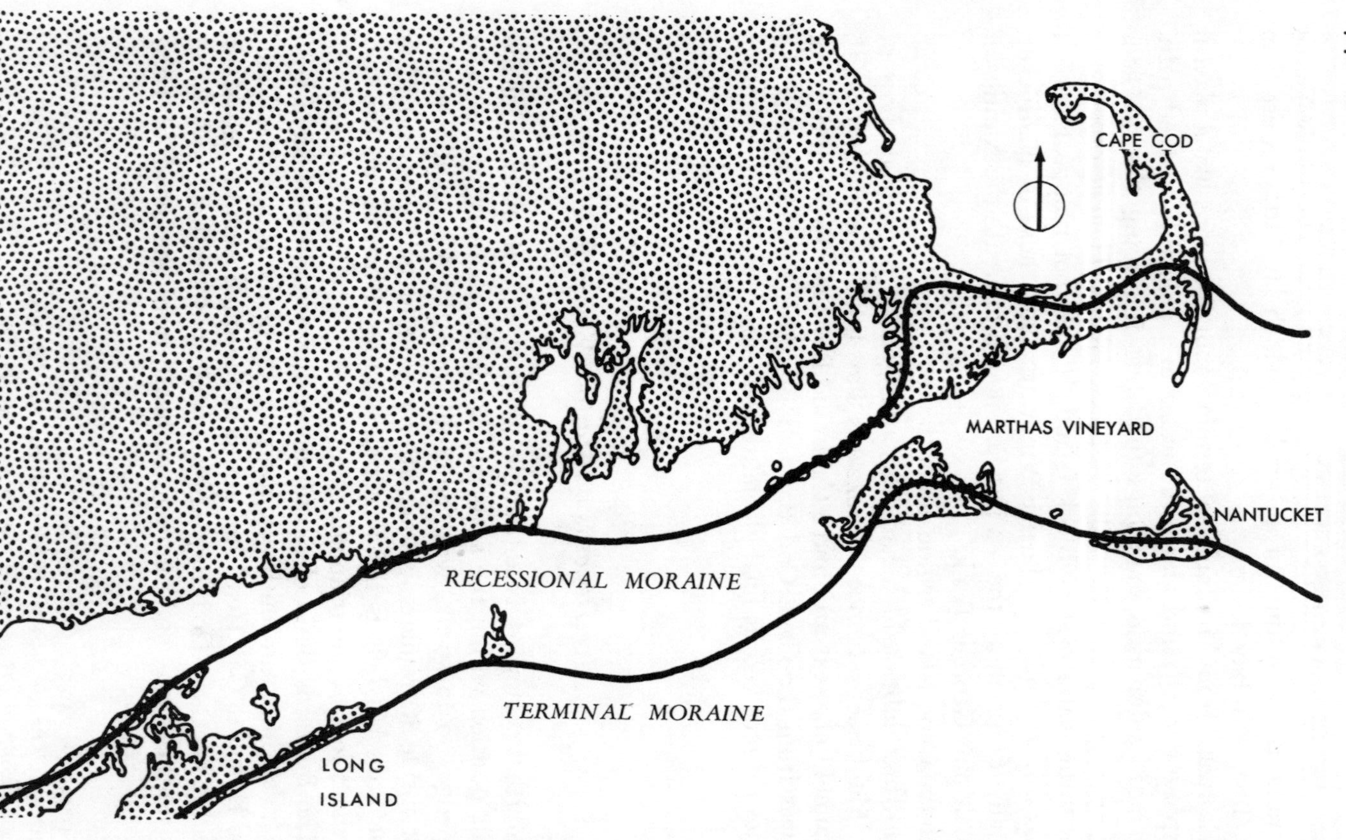

Figure 12. *The moraines across New England and New York.*

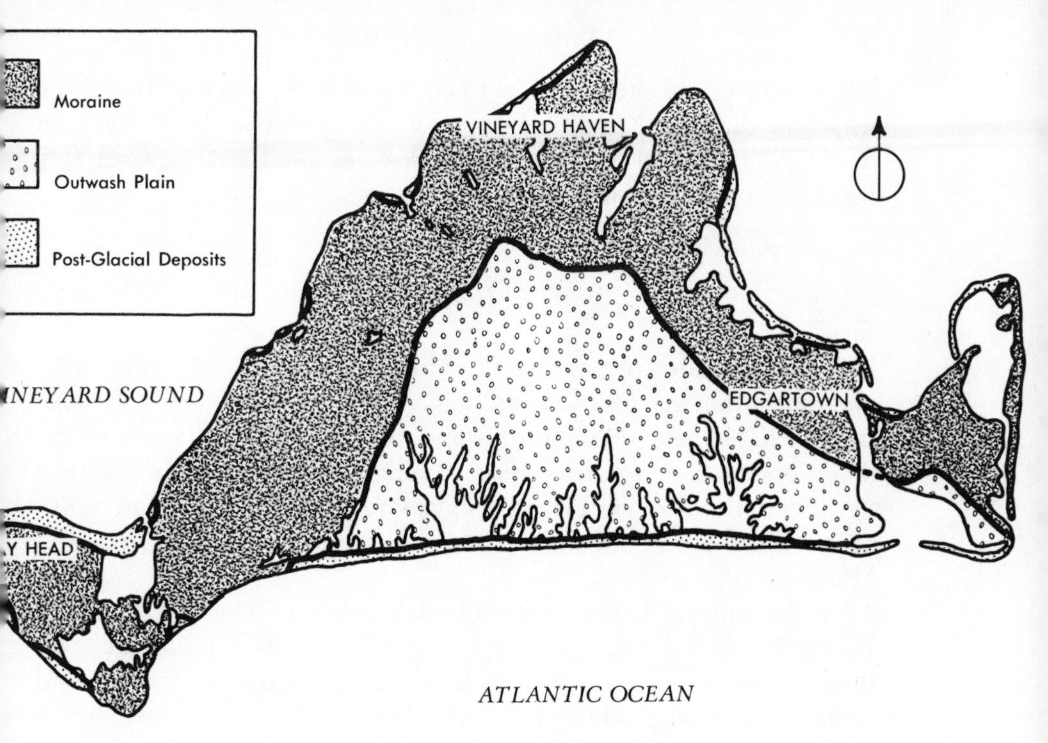

Figure 13. *Geologic map of Marthas Vineyard.*

Tashmoo to Chappaquiddick and thence across to Nantucket's Tuckernuck. Since the moraines meet at Lake Tashmoo, the head of the lake marks the vertex not only of the triangular island but also of the angle formed between two adjoining lobes as they lay in their terminal positions.

The rest of the island is a smaller triangle. It is bounded by the inner edges of the moraines to the northeast and northwest and by South Beach to the south. The bulk of this inner triangle is outwash plain. Surrounding all of the ice deposits, like a decorative border between them and the sea, is a lacy, changing fringe of postglacial sands woven by winds, waves and currents.

The Vineyard Moraines

If you go by boat to Vineyard Haven, you will step off the dock onto an edge of the terminal moraine. You will have to climb uphill on it to get to the center of town. From here, you can travel the length of the island in either direction along highlands. If you turn southeast and head for Chappaquiddick, or southwest toward Gay Head, you will be on a moraine.

The Up-Island Moraine

Southwestward, or up-island, is the Buzzards Bay Moraine (*Figure 13*). From its position we can visualize the towering, debris-laden ice bastion which rose along the island's northwest shore, doubtless broken by crevasses and alive with the flow of meltwaters. From the Vineyard it must have stretched westward just south of the Elizabeth Islands chain and still farther westward nearly to Montauk Point.

On the Vineyard, a combination of Lambert's Cove and Vineyard Haven Roads will take you across a rough, bouldery highland some three miles wide, parallel to shore. Long wrinkles, pronounced dips in the topography, crease the face of the land here, and a spattering of ponds freckles it. Some, as Lambert's Pond, are man-made from dams on small streams. However, most of the blue flecks among the hills, such as James, Seths, and Old House Ponds, are natural, undrained low pockets in the moraine, dammed by ice debris and fed by springs.

The hilly wrinkles of the moraine swell to their largest dimensions in Chilmark (*Figure 17*). Pilot Hill near Lake Tashmoo is only 140 feet high; West Tisbury has 262-foot Indian Hill; but Chilmark claims the highest point on the island, 308-foot Prospect Hill. Not far from it is its twin, Peaked Hill, nearly as high.

The moraine is widest, too, in the Chilmark area. Westward from there it narrows, runs across Nashaquitsa beside Menemsha Pond, and undulates as the wide, wild moors of Gay Head. Beyond the road the moraine passes into the sea at the Nashaquitsa Cliffs.

All along, irregularities in its marginal contours create saltwater ponds such as Lake Tashmoo and Nashaquitsa and Chilmark Ponds.

Superficially the scenic up-island landscape seems a fine example of glacial dumping. However, if we examine the topography closely, a pattern emerges of long northeastward-trending ridges separated by low valleys. The hills are high points on these ridges. One of the valleys contains North Road, with the Peaked Hill ridge on its south and the Menemsha-Prospect Hill ridge on its north.

Thus, the land outline is not haphazard enough to be a typical ice moraine, which would be simply scattered piles of debris. Let us dig down through the glacial drift. We should expect the hills to be built of it. But as we probe we find only three to five feet of this glacier's till in most places, never more than ten feet, and in some places none at all—even at the highest hills. The till is nowhere near as thick as the topography is high. The hills therefore must be expressions of a thickly layered foundation of Cretaceous, Tertiary and earlier Pleistocene deposits. Apparently the Jameco and Manhasset ice advances here folded these earlier deposits high into ridges like corrugated sheets of paper.

Then, when the Buzzards Bay Lobe forced its final way across low and high parts, it blanketed all with its till. Frequently the ridges tore loose boulders from the lobe's lower surface, creating boulder concentrations in the till. Nearly everywhere, stones are abundant. They range from mere pebbles up to twenty-five-foot blocks of granite, clumped in piles, as one geologist described, like "ruined Cyclopean masonry," or simply strewn across the surface.

Long before the arrival of settlers, Indians used these glacial erratics to mark the resting places of their dead. When the white men came and the fields were cleared, the boulders found wider use in foundation work, jetties, and the flat-chiseled stones of walls that run with neat abandon up and down the Chilmark-Tisbury countryside.

If we go to the steep cliffs where waves have cut back the land, from Gay Head to Norton's Point, we see a different kind of till. Here it is as if it has been sieved to remove boulders. Most is loose gravel, contrasting with earlier deposits on the Vineyard so commonly full of clay. This till is faintly layered.

Here we have ground moraine. As the ice moved southward, friction under it removed sand and gravel, leaving these to coat the land surface beneath the ice. Later, when the ice had retreated to lay bare the land, its meltwaters washed this gravel, carried and sorted it and redeposited it in layers.

Many streams cut through the western part of the Vineyard. This makes the region unique on the island, because elsewhere there is so much sand and gravel that it soaks up water like a sponge. Compact preglacial and glacial clay, impervious to water, underlies the western moraine and is close enough to the surface to support the flow of water across the land.

Many of the running waters carve silver swaths across the corrugated surface of the moraine. Some go north to Vineyard Sound, others south to Chilmark Pond. If you follow the moraine up-island you can't help crossing some of these, such as the Tiasquam River, across which Middle Road swings several times.

Other brooks have chosen, as brooks so often will, the most effortless routes along which to flow, and in many places occupy valleys between, rather than across, ice-shoved ridges. One runs from near Peaked Hill to North Tisbury. The gentle sag of another valley parallel to it can be traced all the way from south of Peaked Hill to just beyond West Tisbury, where outwash debris buries it. The Tiasquam River has borrowed this valley for a part of its extent.

Along the Vineyard's northwest shore are smaller brooks, which have taken over channels originally carved by waters flowing from the ice front. A similar channel, submerged and partly dammed by postglacial sandbars, now holds James Pond near Lambert's Cove. In most of these channels, hastening to sea down shoot-the-chute channels, the brooks are more lively than the lazy larger streams to their south. So rapid are they that they were used, during the eighteenth and nineteenth centuries, to power mills. Cornmeal and bricks, cloth and china came into being with their help.

Thus, from the Buzzards Bay Lobe of the scenery-maker, the Vineyard received its most dramatically picturesque landscape. Here are its highest hills, its only streams, and nearly all its small ponds. Even the most casual observer can notice the difference between the up-island terrain and that of the rest of the island.

The Northeast Moraine

The other moraine is that of the Cape Cod Bay Lobe (*Figure 18*). With no high folded sediments billowing beneath it, it has a somewhat more subdued character.

Back at the triangle's apex, at the head of Lake Tashmoo, this Cape Cod Bay Lobe moraine swings from the southeast to meet that of the Buzzards Bay Lobe. Near here the ice lobes joined; and here we can see a clear-cut example of the effect a glacier may have in the shaping of lands.

From an aerial view or on a map we can trace Vineyard Haven Harbor and nearby Lagoon Pond as long fingers of water which thrust southwestward into the Cape Cod Bay Lobe's moraine. Let us imagine the water replaced by fingers of ice. We then can picture two small ice lobes carving their way into the land and leaving West and East Chops to protrude as headlands. These minor lobes must have come traveling the way the valleys runs, southwestward. But they face into the Cape Cod Bay Lobe's moraine. And we know that this lobe traveled generally southeast.

To resolve the seeming discrepancy, let us consider the way an ice sheet travels. As it moves forward, it spreads out in all directions. From a central axis which can be imagined to cut the lobe in two lengthwise, ice radiates outward. The Cape Cod Bay Lobe itself as a whole moved southeast against Nantucket. On the Vineyard, Vineyard Haven is nearly at the western edge of the wide lobe. Thus, in this region the irregular advancing margin of the lobe simply pushed to the southwest. Small tongues spread out beyond the western margin and likewise moved southwest. On their paths they scooped out the valleys. Later, after the ice had gone, rising seas were to submerge these valleys and throw up the sandbar, crossed by Beach Road, which now nearly separates them.

From Vineyard Haven the Cape Cod Bay Lobe's moraine runs southeast to form the island's eastern side. It narrows slightly toward the southeast corner before cutting across the northern neck of Katama Bay at Edgartown to Chappaquiddick's Long Point. Pleasant and gently rolling is this moraine, but not topographically spectacular (*Figure 18*). Unlike the 300-foot summits in the moraine to its west, the highest hilltops here reach scarcely higher than 120 feet; these are in Oak Bluffs. Ninety-four-foot Sampson's Hill is the highest point on Chappaquiddick. Boulders are much scarcer here than in the western moraine. And as we have seen, the paucity of clay here allows no streams and few ponds. Those ponds which do exist lie for the most part along shores. As irregular low areas in the edge of the moraine, they have been filled by the sea and cut off by sandbars. Such are Crystal Lake at East Chop, Farm Pond at Oak Bluffs, and Trapps Pond at Edgartown.

However different the scenery may look from the other moraine, the till looks the same. Typical glacial sand, gravel and boulders lie within the Cape Cod Bay Lobe's moraine. Directly along the shore, at Lagoon Pond's banks and around Vineyard Sound and high in the East Chop cliffs, we can see gravelly ground moraine like that across the island at Nashaquitsa. It is mostly small pebbles of granite, less than two inches across, from the mainland.

The Southern Edge of the Ice on the Vineyard

Because both of the island's moraines are parts of the terminal moraine, their inner, southern edges represent the farthest extent of the scenery-maker. Like the line of trenches at a battlefront, a trough marks the front edge of the glacier's journey. This trough is shallow, vague and disconnected, with boulders strewn in it. We can trace it locally where it runs just along the southern edges of the moraines (*Figure 13*). On the west we pick it up best about a mile north of West Tisbury; and we can follow it from here northerly along the inner margin of the hill belt. On the east we find it again on Chappaquiddick. Here it is harder to see but is marked by a telltale line of boulders, some twelve feet across. Starting about a mile and a quarter northwest of the lower end of Poucha Pond, we can follow it across south-central Chappaquiddick, swing with it west and then northwest, where it skirts Sampson's Hill on its south. Naturally, boulders have been tampered with in many places. We lose the trail temporarily after crossing Katama Bay until just south of Lagoon Pond. Here we find it again clearly, heading west-northwest for several miles and containing Duarte Pond in its deepest part.

Wherever this boulder-strewn trough runs, a moraine walls it on its north and an outwash plain rises from it on its south (*Figure 14*). For when ice lobes lingered in their terminal positions, meltwater streams flowed from them with sand and gravel. Some distance in front of the glacier they slowed up, building wide fans of their alluvium. This left a narrow area between moraine and alluvium almost without deposits, excepting boulders too heavy for the streams to move far.

The Vineyard's Great Plains

Starting at Edgartown and traveling west to West Tisbury, the Takemmy Trail takes us from moraine to moraine via the island's

Figure 14. *The edge of the ice lay here. This bouldery trough lies just beyond the former ice edge on western Marthas Vineyard. Ice lay to the right and piled up morainal debris; meltwaters flowed to the left, building the outwash plain and releasing their largest rocks at the foot of the ice.*

U. S. GEOLOGICAL SURVEY

Great Plains (*Figure 15*). These flat sandy plains much impressed geologist Hitchcock when he first viewed this region in 1824, before the glacial theory was known. Gazing across the then-barren expanse of sandy fields, he hesitantly viewed the plain as the "retired bosom" of sands eroded by the sea from the island's hills. Not long after, enlightened by the work of Agassiz, he found it to be outwash plain.

Today pitch pines, white pines, scrub oaks and wild flowers cover the plain. It is not heavily populated with people, although cleared portions indicate that dairying and truck farming are taking hold. In the shadowed recesses of the yet-wild parts live many forms of wildlife. The 4500-acre Marthas Vineyard State Forest in its center provides protection. Here the gentle heath hen made its last stand as a species until 1933, when forest fires and shotguns erased the last individuals from the earth.

No streams survive on this porous plain. Beneath the surface soil there is clean, rather well-sorted sand and gravel. If you wanted to dig through this to see what is beneath, you could keep digging until you had gone as far down as a five-story building, and still you would be in sand and gravel.

Actually it is nearly impossible to define a line of division separating this outwash from earlier sandy deposits on which it lies. Since the outwash materials came from the ice, and since this scraped much of its load from such local sandy deposits, there is little difference between the two. As nearly as can be judged, however, there are some fifty feet of actual outwash beneath the plain.

Boulders are rare in any outwash plain. The moraine at Gay Head supplied the largest boulder in the Mayhew Monument, at the Place by the Wayside on the Takemmy Trail. The smaller cob-

Figure 15. *The Great Plains of Marthas Vineyard. This shows the typically featureless outwash plain with the western moraine in the distance.*

U. S. GEOLOGICAL SURVEY

bles piled up here by Indians doubtless were gathered locally. They are about the size limit of outwash rocks.

The Vineyard's Great Plains are very nearly featureless (*Figure 15*). Both the Marthas Vineyard airport and the Edgartown airport are located here, just as Cape Cod's Hyannis airport and Nantucket's airport also lie on outwash plains.

Actually, the surface is not level. Like the surface of any alluvial plain, we can expect it to be higher near the source of the alluvium than away from it. Thus, its highest elevation, about a hundred feet, is where the plain borders the moraine. From here it slopes about twenty feet per mile southward to the shore of the Atlantic along South Beach.

When you travel by car along the Takemmy Trail the road may seem nearly flat, but should you go under your own power, by bicycle for instance, you soon will be aware that it crosses a series of gentle valleys or grooves, called "bottoms." These start halfway across the plain and, a mile or less apart, trend at right angles to the road.

Such bottoms hold in their southern parts the most interesting features of the plain, many branching, stringer-like ponds. The two largest, Edgartown Great and Tisbury Great Ponds, are composite, with their branches, or "coves," fanning out northward like the feathers of a grouse's tail. Others—Chilmark Pond, Quenames Cove and Black Point Pond in Chilmark; Homer and Watcha Ponds in West Tisbury; Oyster, Paqua and Job's Neck Ponds in Edgartown —are simply long fingers reaching out to touch the southern edge of the island. Skirted by the road on their northern edges, they provide the openings for frequent and provocative seaward vistas.

The grooves deepen and widen toward the south. Apparently they were the channels of onetime streams. The size of these streams reflected the tremendous influx of water from the melting ice. When conditions became drier and sands blotted up their waters, the streams vanished.

In the coves today the ponds lie at sea level, in the wide mouths of the extinct glacial streams. For as the ice vanished, seawaters filled these stream mouths and turned them into little estuaries and bays. Then the ocean built a barrier beach, South Beach, walling off the bays from the ocean. Gradually their waters freshened and they became ponds. Katama Bay between Edgartown and Chappaquiddick had a similar origin, but so far has kept its opening to the sea.

Thus, Marthas Vineyard is almost entirely an ice-molded island. Its rolling hills and valleys, its great plain, its stone walls, blue coves and refreshing scenery—even its shape—it owes to the glaciers. The only changes here since the days when ice first exposed the Vineyard to the air have been wrought about its shores where the sea has played, taking land from here and tacking it on there —and across its surface, onto which winds and rainwaters, vegetation and people have brushed the colors and shadings of a landscape.

The Ice on Nantucket

The terminal moraine of the Cape Cod Bay Lobe, as we have seen, drops beneath the sea at Chappaquiddick. Five miles east of the Vineyard it shows up again on Tuckernuck Island. A bit farther east it rises as part of Nantucket Island itself. Southeast of Nantucket it remains hidden to create the dangerous Nantucket Shoals.

"The island of Nantucket," said J. B. Woodworth, ". . . is one of the most instructive portions of the terminal moraine of the last ice epoch in North America, because it is the most distinct and isolated of the glacial accumulations. Set in the waters of the ocean far to the south of the morainal belt of Cape Cod, and distant nearly its own length from the neighboring island of Marthas Vineyard, the peculiarities of its glacial form, despite the low relief of the island, are readily discerned."

And the relief really is low. The moraine here is simply an open, gently undulating surface running along the northern part of the island (*Figures 16; 19*). Its building materials and terrain resemble closely those of its near relative, the Cape Cod Bay Lobe's moraine on the Vineyard, although the relief on Nantucket is even less pronounced.

The southern bulge of Nantucket's thick crescent is its outwash plain; and surrounding both moraine and plain are the sandy constructions of the sea.

Nantucket's Moraine

Arriving at Nantucket Island by boat and docking at the harbor, we see the very low edge of the moraine on either side of the harbor.

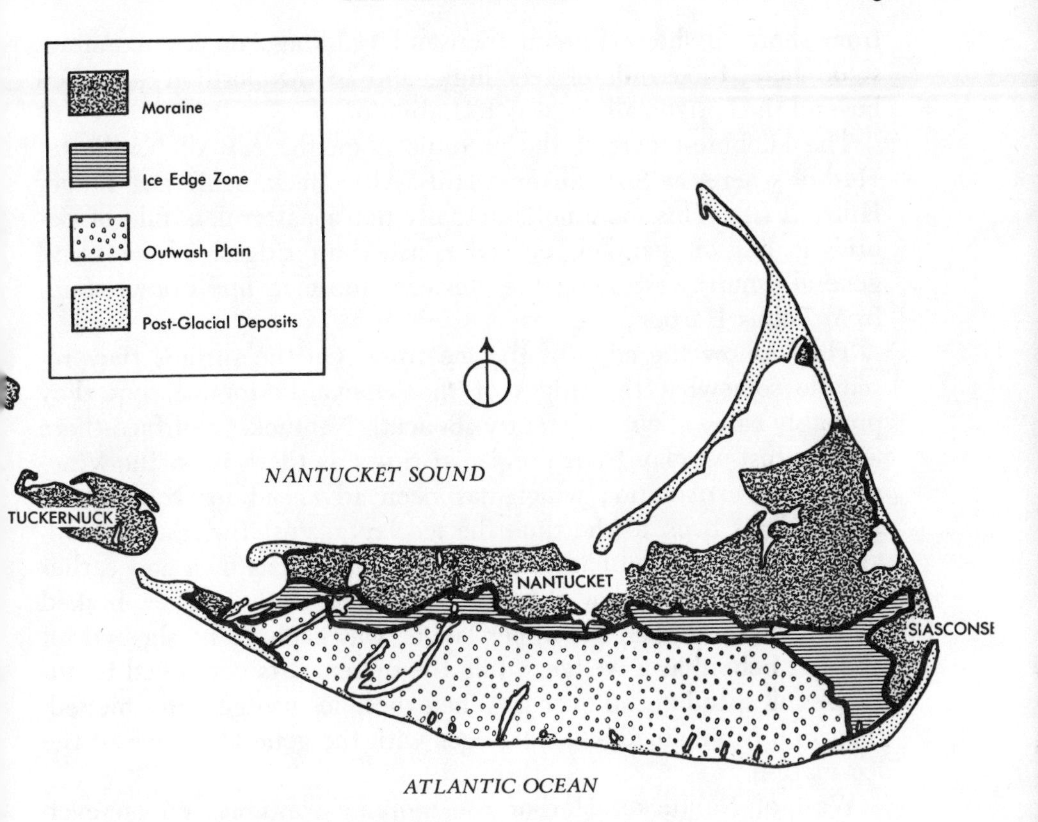

Figure 16. *Geologic map of Nantucket.*

As soon as we leave the dock we find ourselves on its hills, for the ups and downs of the old town's streets follow the subdued morainal contours.

In the southern part of Nantucket Town, gentle hills are especially common—the watchtowers, landmarks and wind catchers of the town. During Nantucket's golden whaling era, women and children frequently climbed these hills to watch for a special ship's return. In 1686 Nantucket's first house, the Jethro Coffin House, went up on one such hill, now called Sunset Hill. The old windmill in Mill Hill Park, built in 1694, is the last of four such mills which once rose from morainal hills about town to catch the sea winds.

Beyond the town, Polpis Road takes us eastward along the moraine at close range, the Shawaukemo Hills rising on our left. From the higher summits we can look across the sea-bordered landscape from shore to shore. Fifteen thousand years ago an ice mountain stood here, thousands of feet high, and to the north it stretched beyond the horizon all the way to Labrador.

The knobbiest part of the moraine is on the east of Nantucket Harbor where the Shawaukemo Hills, Altar Rock, Saul's and Folger Hills all rise. This moraine is actually not a patternless pile of debris; it lies on parallel, curved, pushed-up ridges. These trend generally northwest along the western side of a line drawn south from Polpis Harbor.

They follow the edge of the ice front. On the surface they resemble somewhat the ridges of the Vineyard Moraine, but they probably came about differently. Beneath Nantucket's surface there is no thrust-up clay from preglacial times as there is on the Vineyard. One explanation which has been advanced for Nantucket's ridges dates back to the time the ice lay against the island. Probably it now and then had to flow over banks of its own and earlier sandy outwash. In doing this, the lower part of the glacier, braked by friction more than the upper layers, was likely to be sheared off and left behind as a wedge, while the upper parts continued to advance full speed ahead over it. When these ice wedges later melted, their debris remained as long ridges with the general outline of the ice margin.

West of Nantucket Harbor the moraine continues on an even more gentle scale. We can take Cliff Pond Road or Maddaket Road from the town and travel between the miniature Trots Hills, which rise no more than sixty-five feet. On the west at Maddaket Harbor, the hills drop to twenty feet by the time they reach the sea. Beyond the shore the moraine breaks through the water surface as the northern rim of Tuckernuck Island, and still farther west, as the northern half of Muskeget Island.

Clay is rare in the sandy till of this moraine, although in places enough was found to use in road building and in making bricks for Nantucket buildings during the eighteenth century. In some places there are pebbles of red, fossil-filled Cretaceous sand, and rock fragments from the mainland.

None of the till is thick. In only about three spades' depth we shall have passed through it completely. Then comes layered sand and gravel outwash of the earlier Manhasset glaciation, when the ice itself never reached this far south.

Beneath Sankaty Head's ocean cliffs, the till materials lie in cross

Figure 17. *Morainal landscape on western Marthas Vineyard. This landscape was created by the Buzzards Bay Lobe at Chilmark; the stones in the wall are from morainal debris. Compare this with the landscape in Figure 18.*

U. S. GEOLOGICAL SURVEY

Figure 18. *Morainal landscape on eastern Marthas Vineyard. This landscape was created by the Cape Cod Bay Lobe, near Sengekontacket Pond. Compare this with the landscape in Figure 17.*

U. S. GEOLOGICAL SURVEY

Figure 19. *Morainal landscape on Nantucket. This subdued landscape was created by the Cape Cod Bay Lobe across northern Nantucket; it is nearly lacking in boulders. Compare with Figures 17 and 18.*

U. S. GEOLOGICAL SURVEY

section above the Sankaty Sand. Here thin, water-sorted layers of gravel, reworked from the actual ice deposits, reveal how little they resemble true till.

Hidden between ridges and scattered among hills are small basins. These are relatively low areas which happened to be surrounded by ice debris. Springs have filled some of them and turned them into ponds. In time, these ponds have tended to fill with vegetation, to become swamps and finally to dry out; we spot their former locations by the patches of lighter green swamp plants which frame them.

Despite its rolling topography, Nantucket's glacial terrain is incomplete. Scarcely a boulder appears, though we tramp the length and breadth of the island—although a few cobbles show up here and there. As we have seen, large rocks also are somewhat scarce in the Vineyard part of the same Cape Cod Bay Lobe's moraine. Doubtless, therefore, they were never abundant on Nantucket. Yet there is evidence of a former rock supply larger than today's. The old Indian name for the town of Sherbourne, ancestor of today's Nantucket Town, was "Wesko," or "White Stone." By the end of the eighteenth century this apparently impressive light-colored stone lay concealed by the wharf in the harbor.

Countless other boulders have vanished from fields and hills. In the early days of settlement, these rocks were the only natural building materials available; there was no bedrock building stone, wood was scarce, and although there was some clay, not enough was found to fill local demands.

For more than two and a half centuries all sorts of masonry, wharf and foundation construction across the island have taken tons of the glacier's stony cargo. Even today you can still see the holes which pit the island's surface in places from which rocks were removed. In fact, in a few places where boulders once were numerous, the needs of road building alone all but destroyed the moraine itself by stripping it of its rocky substance. Geologist Shaler remarked that the area was "more thoroughly stripped of its boulders than any other [region] known to me." In fact, even back in the first part of the eighteenth century, so few boulders remained, wrote Crevecoeur in his idyllic description of Nantucket, that stones had to be imported for wharves and cellars. And every stone of the cobblestone pavement newly sported by Nantucket Town's Main Street in 1837 came as ship ballast from the shingle beaches of Gloucester.

Nantucket's Moors

Milestone Road takes us across Nantucket's moors from Siasconset to Nantucket Town (*Figure 16*). This is the flattest part of the island. It is, of course, outwash plain—a desert of sand but abloom with wild flowers: broom; plum vines; arbutus; sweet fern; daisy; azalea; heather and heath; elderberry and wild rose; pitch pines and scrub oaks. Low-growing, these plants hug the ground and avoid the force of the sea winds which sweep across the plain.

At one time the moors were known as the "commons" and used to pasture sheep, a thriving industry on the island. In time, the low cropping of plants by the grazing sheep so destroyed vegetation that the fields had to be retired as unprofitable. Today they shelter the small native deer, hosts of smaller mammals and a rich variety of birds.

Beneath their green and flowering cover, the moors consist of some twenty-five vertical feet of outwash sands and gravels. Clay is rare indeed. Because of this it would take some imagination to call anything on this part of Nantucket a stream. A few ill-defined valleys connect ponds here and there, but water scarcely moves along them.

Traveling across the outwash plain takes us across a series of parallel creases which run slightly northeast-southwest. Some reach from the island's southern coast to all the way across the moraine on the north. Madequecham Valley, east of the airport, is one of these, cutting across the moraine west of Altar Rock as far as the coast. The two wings of the Hummock Pond channel farther west reach northward to touch the moraine; the eastern branch of this depression can be traced to the north shore, cupping a chain of ponds on the way. Most of the other channels reach only about two-thirds of the way across the moors to the island's northern shore, many harboring ponds at their southern ends.

One of these is saltwater Hither Creek (Maddaket Ditch), southeast of Maddaket Harbor at the western tip of the island. Tides fill this now and make it an important herring run in the spring.

To its east, Long Pond heads in North Head Long Pond, which drains by a narrow creek into Long Pond itself. Sheep Pond, south of the Homing Facility Station, lies in the southern tip of a short crease. East of large Hummock Pond is little Reedy Pond, heading in the Larraby Swamp east of Hummock Pond Road. Great Miaxes Pond comes next, then Miacomet Pond, branching on its north into two parts like a Y. The Weeweeder Ponds near Surfside are now swamps; eastward from here the valleys are empty and some are nameless; Nobadeer Valley and Forked Pond Valley are the largest. Tom Nevers Pond is connected with Gibbs Pond along a valley occupied by Phillips Run, one of the few streams on the island.

All of these are analogous to ponds along the Vineyard's southern shore. Their valleys first were carved by meltwater streams, then drowned at the mouths by rising seas and finally dammed by sandbars. Since sea level was lower when the pond beds formed, the bottoms of some are below sea level today. Hummock and Long Ponds fill the deepest of these.

Walking on the beach along Nantucket's southern shore, we pass low bluffs which increase in height as we go eastward to as far as Tom Nevers Head. Originally the fan-shaped sheets of outwash deposits of the moors sloped gently south to their termination in the point of a gradual wedge. The present truncation of the wedge by sea cliffs here shows that the southern tip of it has been removed by the sea. From the southward slope of the plain and the height of the bluffs we can calculate that the wedge of outwash extended originally a mile or more south of Nantucket's present southern coast. Working on these loose deposits, the sea has encroached this far into the original island.

The Southern Edge of the Ice on Nantucket

Nantucket, like the Vineyard, has a belt of lowland representing the actual location of the southern limit of ice (*Figure 16*). Near Gibbs Pond is a giant cranberry bog carved out of a swampy belt. This swamp, twenty to forty feet above sea level, runs along the northern boundary of the moors. It is half a mile to a mile wide and is the flattest surface on the entire island. Three of Nantucket's four major swamps lie here: the boggy land in the west near Maddaket Harbor; the cranberry bog south of Saul's Hills; and Tom Nevers Swamp near the 'Sconset golf course.[7]

Bordering the northern edge of this very flat swampland is the moraine; bordering its southern edge is a twenty to forty foot rise in the land up to the level of the outwash plain. This rise is most noticeable south of Saul's Hills in the vicinity of the cranberry bog; it becomes indistinguishable near Nantucket Town, but can be picked up again farther west near Hummock Pond.

Along this poorly drained belt lay the front edge of the Cape Cod Bay Lobe. South of the ice, meltwater streams built up the moors. In the low area itself there was little deposition except for some thin-washed sandy drift and an occasional boulder. If you make your way across the swampland surface, therefore, you may

[7] Squam Swamp, the fourth large swamp belt, lies not in this region, but in a low, undrained area within the moraine.

be treading on the actual unglaciated surface of an ancestral Nantucket Island.

Tuckernuck and Muskeget Islands consist primarily of outwash, with only a thin band of higher moraine across their northern parts. As for the rest of Nantucket, only Coatue and Coskata Beaches and a few narrow shoreline areas came into being after the ice left and winds and waves took over the sculpturing job.

The Moraine Beneath the Sea

Between and around these ice-built islands, wide areas of sea floor are corrugated with shoals and dangerous to ships. On old sailing charts of the Sounds and Buzzards Bay, groups of rocks on some of these shallows are given such fanciful names as Rose and Crown; Hen and Chickens; Bishop and Clerks. The boulder-topped shoals still bear these names (*Figure 29*).

They are parts of the vast submarine deposits of the Buzzards Bay and Cape Cod Bay Lobes, for naturally not all of the terminal deposits of these ice bodies rise as islands. Between and around the present islands, beyond the reach of men, the boulders of the submerged moraine have remained to harass sailors.

Even relatively far at sea there are submerged glacial hills. Off Nantucket some twenty-three miles east and forty miles south of Sankaty Head are the Nantucket Shoals, marked by the Nantucket lightship. According to a recent *Atlantic Coast Pilot,* they constitute "one of the most dangerous parts of the coast of the United States for the navigator." In some places, the tops of these shoals lie only three or four feet beneath the surface, and heavy breakers crash over them. There are boulders here too; an old chart of the region refers, with Yankee brevity, to one part of the shoals: "Bow Bell is Great Stones."

The actual distance which the Cape Cod Bay Lobe stretched seaward from Nantucket is unknown, but the material topping the shoals apparently is one of its farthest deposits. It is certainly possible to imagine the ice extending this far over what is now water. Sea level was lower by some 450 feet during the glaciation. The ice must have been several thousand feet thick. It would have scraped easily across the shallowly submerged continental shelf for many miles beyond Nantucket. In its terminal position, it must have lain against the shoals, which then rose as part of the old coastal-plain ridge, pushing debris upon them to add to their treachery.

Still farther east, another part of the drowned coastal-plain ridge, Georges Bank, likewise was covered by part of the terminal moraine. Georges Bank is probably the work of the South Channel Lobe which adjoined the Cape Cod Bay Lobe somewhere east of Nantucket. The ice of this offshore lobe etched the Bank into sharper relief, scooping out the lowlands north of it and grinding to a halt against the old ridge to pile up a moraine.

Ice debris covers the top of Georges Bank. Fishermen who work the Bank have hoisted rocks weighing as much as three tons, entangled in their drag nets by coralline growths. Now and then they snag still larger boulders which break the nets.

Early reports of such rock encounters went to the Fish Commission in Gloucester which, in the late nineteenth century, "stimulated a worthy rivalry among the captains and fishermen to bring in every thing zoological or geological which might possibly be of scientific interest for the Fish Commission," as Upham noted. These reports were of scientific interest to geologists too, for from them was acquired the first information about the glacial nature of the tops of the Bank—and with this, valuable clues as to the extent of the ice and the history of the entire region.

The top of the Bank, like that of the Nantucket Shoals, is brought dangerously close to the sea's surface by the moraine, within twelve feet in several places. Storm-driven seas pile up on these shallows to create a giant surf. The government's former Texas Tower installation stood rooted on Georges Bank on sixty-foot legs, yet storm waves have reared high enough to pound the tower platform. Combinations of winds and the fierce run of the tides frequently cause still shallower conditions, in which only a few feet of water cover parts of Georges Bank; indeed, there have been reports of the emergence of dry land. Thoreau wrote:

> Every Cape man has a theory about Georges Bank having been an island once, and in their accounts they gradually reduce the shallowness from six, five, four, two fathoms, to somebody's confident assertion that he has seen a mackerel-gull sitting on a piece of dry land there. . . . There must be something monstrous, methinks, in a vision of the sea bottom from over some bank a thousand miles from the shore, more awful than its imagined bottomlessness; a drowned continent, all livid and frothing at the nostrils, like the body of a drowned man, which is better sunk deep than near the surface.

Apparently fishermen of the Cape and Islands thought likewise. Although one ship's crew in 1796 is said to have played ball upon it, most eighteenth-century sailors gave this almost legendary land a wide berth. But the dangerous currents which make the Bank treacherous also make it valuable; for they sweep animal and vegetable matter into the region, providing food which encourages fish to congregate. Through the fog of fears and legends of the eighteenth century, the Bank's vast schools of cod and halibut were pulling fishermen like magnets. Finally and inexorably, in the nineteenth century fishermen were drawn to these backyard gold mines—for if the risk was great, so were the profits. For more than a century since, Georges Bank has fed and clothed many Cape and Island families—and bereaved many, too.[8]

[8] Fifty-seven Truro men and boys were lost in a single storm on the Bank in 1841.

CHAPTER VI

The Scenery-Maker

II–CAPE COD

On Cape Cod one glacial force shaped many landscapes. . . . Cape Cod is a well-illustrated chapter in geologic history, setting forth the story of glacial deposition in gigantic but simple form. . . . [Cape Cod's] Great Beach and its related hinterlands possess varied natural qualities of national significance. . . . There are many square miles of spectacular dunes, fresh-water ponds, picturesque, rolling heathlands, salt and fresh-water marshes, ancient river valleys, [and] forests. . . . The combination of these features cannot be matched anywhere on the Atlantic coast.

U. S. NATIONAL PARK SERVICE
Report on a Cape Cod National Seashore

The Retreat of the Ice

Like a mighty creature at bay before the warming air, the ice lingered with its front against the Islands. The Pleistocene Epoch was drawing to a close. Slowly the glaciers began to melt back, leaving southern New England for the last time. They probably released about a mile of land every ten years.

As the ice wasted northward, it bared a widening scene of desolation. Until slowly spreading forests clothed the naked landscape, the only moving things upon it were debris-glutted meltwater streams, windblown sands and, doubtless, some bird and animal visitors. Much of the material carried by ice-fed streams now lies on the floors of the Sounds.

Although the ice was on its way to oblivion, its journey was indecisive and ragged. Several times it drew to a halt, gathering strength to build new moraines. At least once it achieved a sufficient burst of energy to move forward. It was this final forward movement which formed most of Cape Cod.

Cape Cod's Glacial Landscape

There are four major elements in Cape Cod's glacial landscape (*Figure 20*). Two long recessional moraines block out the Upper Cape, as the east-west part of the peninsula is known. One of these moraines runs north-south along Buzzards Bay from Bourne to Woods Hole; the other runs east-west not far from the Cape Cod Bay shore from Sagamore to Orleans and Chatham. Fenced in by these ridges on its west, north and east, and by the Sounds on its south, is Cape Cod's great outwash plain. The Lower Cape—the north-south forearm—is largely a third moraine, not an ordinary frontal moraine but an interlobate moraine. It consists of deposits contributed by two adjacent lobes and partly reworked and stratified by waters after the ice had left. The rest of the Cape has been built by the sea.

Cape Cod's Outwash Plain

The outwash plain was the firstborn member of the Cape's glacial landscape. While ice moved to and from the Islands, Cape Cod acquired a disorderly array of ground moraine and outwash gravel. It is on these foundations that the present outwash plain lies.

The plain reaches about two-thirds across the Upper Cape from the Sound shores (*Figure 20*). A rim of moraine overlaps it on its north and west and partly on its east. Mile after mile of level pitch-pine forests stretch across this plain, cleared in part for Cape Cod's airport at Hyannis and for Camp Edwards and Otis Air Force Base north of Falmouth. In many places the forests have yielded to vegetable and strawberry farms and to home and cottage construction. Unlike the outwash plains of the Islands, Cape Cod's plain, especially along the Sounds, is a scene of lively activity centered in such towns as Falmouth, Hyannis, Dennisport, Osterville, Cotuit and Harwichport.

Also unlike the Islands' plains are the hundreds of lakes and

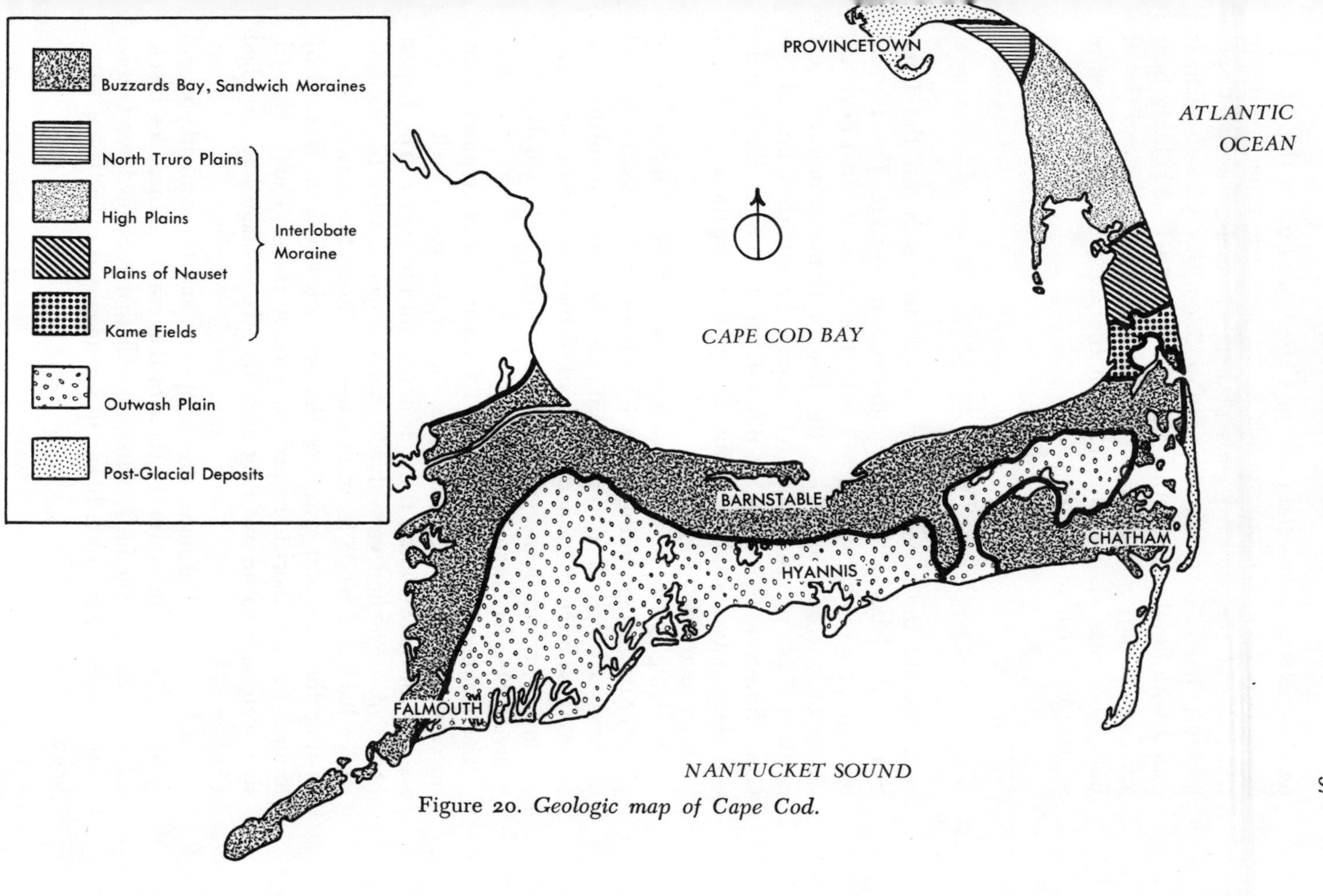

Figure 20. *Geologic map of Cape Cod.*

ponds scattered across the plain of Cape Cod. For on the Islands, as we have seen, the only ponds in the outwash plains are those which lie at the seaward ends of long channels. On the Cape, however, ponds are everywhere. In addition, the outwash plain is pitted with hundreds of other hollows, either dry or holding bays and some of the best harbors on the Cape, such as Lewis and Cotuit Bays. So numerous are the holes in the surface that the entire region can be called a pitted plain.

Origin of the Ponds

Apparently the scenery-maker acted differently on the Cape than on the Islands. And the difference is obvious. The Islands' outwash plains never were covered by the ice on its final advance, which took it only as far as the line of terminal moraines just north of those plains. Completely across the east-west arm of the Cape, however, the same ice plowed, on its way to the Islands, and the front crossed the same Upper Cape again as it melted back on its retreat.

The dying glacier was weak as it shrank northward across bumpy plains of its own ground moraine. Chunks of ice, trapped by ground unevenness, broke off from the main mass. Across the desolate landscape, these huge, dirty derelicts marked the wake of the vanishing ice like grounded icebergs. Today there are placid ponds where many of these ice fragments once lay.

The making of a pond bed out of an ice block required, first, material to build up about the ice; second, the ice to melt, leaving a cavity in the material; third, water to fill the cavity. The largest blocks of ice melted very slowly. Some lingered after the ice had drawn back and while its waters were building the outwash plain. In time the sand and gravel of the outwash alluvium, banked up around the ice blocks, helped to protect them. Thus insulated, they were able to survive long after the glacier itself had vanished (*Figure 21*).

Finally, when the inevitable happened and the sunbathed land surface became too warm for their further survival, these ice blocks followed the glacier into extinction. All that remained were holes in the sandy plain. Such holes, to a geologist, are kettles or kettle holes.

Figure 21. *The formation of a kettle hole. Isolated ice blocks become covered with the pebbly debris of an outwash plain.* U. S. GEOLOGICAL SURVEY

Many have clumps of till on their rims, part of the debris which had been incorporated in the ice before the blocks broke off and which, when the ice melted, was left on the ground moraine.

The bottoms of many of these pits lie far under the surface of the outwash plain. Some go down a hundred feet. Therefore many lie below the water table, and groundwater has flowed in to fill them, creating ponds. In most of these a natural circulation keeps the water clear and fresh; water loss by evaporation is balanced by contributions from underground springs. In this respect, the ponds are really like large, open, natural wells, unlike most inland lakes, which are fed and drained mainly by streams.

If you note the arrangement of the kettle holes on a map or from an airplane, an interesting pattern emerges. Some are grouped in chains. For instance, Mystic Lake, Middle Pond and Hamblin Pond, near Pondsville, are strung out in such a chain; Peters, Wakeby and Mashpee Ponds, north of Mashpee, make up another; and west of Farmersville, Spectacle, Lawrence, Triangle and Hog Ponds are a third.

The peculiar arrangement, like the ponds themselves, must date far back to the Cape's icy past. Perhaps some of the isolated ice blocks at that time were long ice ridges. Such ridges can and often do break from a glacier where crevasses, formed from tension, weaken the ice perpendicular to its front. Earlier, when the front of the Cape ice lay some thirty miles beyond, snugged against Nantucket, such crevasses might have grooved it. Within these cracks, melted ice water would have flowed, slowly widening the splits. At the same time, the flowing water would have left debris in the breaks. Finally, bars of ice would have become completely isolated from the main mass, to be left behind when the ice re-treated. Later, many would have broken further along transverse weak zones. The smaller chunks, banked over by more debris, finally would melt completely, leaving the chains of pond beds as they exist today.

The widely varying shapes and sizes of the kettles across the plain reflect their haphazard beginnings. There are some which are not steep enough even to retain water and are dry depressions in the land surface. The sides of others slope five to ten degrees; still others drop twice as much, putting you in water over your head by the time you have waded fifteen feet from shore.

Some ponds are even split-level. The water lies in a lower part beneath a sand-covered terrace which is rimmed by the still higher ground around the pond. Wakeby Pond, just north of Mashpee Pond, is one of these. The pond is cliffed on its western shore, and a sandy, flat bank forms its northern and eastern sides and its northeast corner. The upper part of the ice block must have melted somewhat while outwash was still spreading about it to its west. The melting fragment would have spread some of its own sand and gravel across the northeastern rim of the pit. Meanwhile, its very presence there prevented the outwash gravel from piling up to the level of the upper rim. Thus, a terrace remained at the foot of the rim.

The South Sea

During all this outwash plain construction, the ice remained split into three lobes. One lay north of the present Upper Cape shore, stretching across what is now Cape Cod Bay. Another reached down across Buzzards Bay alongside Cape Cod's present Buzzards Bay shoreline. The third was farther east. In these positions the ice lingered, possibly for several centuries. Gradually its broad fans of outwash sand coalesced as they widened to form the continuous plain as it is today.

Sloping seaward toward Nantucket and Vineyard Sounds, the plain drops twelve to twenty-five feet per mile. In a few places, especially around Sandwich and Falmouth, it is four times as steep.

Early Cape Codders called this plain the "South Sea." Towns clustered at first along the moraines and bays, and the sandy expanse remained remote and uninhabited. Then the discovery of shellfish in the Sounds drew people across the plain, and towns grew up on the sea of sand.

It is a deep sand sea. In some places the sand and gravel reach as far beneath the surface as North Truro's cliffs rise above the sea. Tracing these deposits northward and westward, we find that when we reach the bulky moraines along the shores, the gravel seems to disappear beneath their till. Thus we know that the plain was built up first and the moraines were later dumped upon it. If we follow the gravels southward, we see them end as very low bluffs at the edge of the Sounds. Beneath these bluffs is more gravel, extending under the sand of the beaches. And under Nantucket and Vineyard Sounds themselves, down beneath the water and the sandy sea bottom, outwash plain debris reaches from half a mile to a mile farther than the Cape's southern shore.

Like strangers out of place among all the gravel of the plain, there are patches of bouldery till, ground moraine left from the glacier's southward journey and by chance not covered by outwash. Stone walls ramble across small hills rising up from the Cape's flat lands. They alone spell till, for no such large stones occur among ordinary outwash materials. Cleared from field surfaces or turned up with shovels from red till, these stones are particularly common around Hyannis and Hyannis Port. And many patches of ground moraine have survived along the coast. They cap Harbor Bluffs on the shore of Lewis Bay, and partly cover Craigville and Great Island, especially Point Gammon. East of Lewis Bay, and still farther east near Dennisport, there are more. To the north, throughout the interior of the outwash plain, tiny till patches, only a foot or two across and less than a foot thick, lie like miniature islands in the South Sea.

Channels in the Plain

Cutting into the land as if gouged by a giant chisel, broad shallow furrows appear in the southwest part of the outwash plain (*Figure 22*). They start about halfway across the plain and reach to its southern shore, nearly at right angles to it. In cross section they are V-shaped. But so broad and shallow are they that at their mouths some reach a thousand feet across but are only twenty feet deep. Although especially evident from Falmouth eastward to Succonesset, such valleys can be found still farther east across the entire southern edge of the plain to its end near Chatham.

A series of the channels passes beneath the shore road which goes eastward from Falmouth Heights to Menauhant. Long "necks" of land separate them. As the sea rose following the end of glaciation, their mouths became marine inlets, creating harbors. Today, however, only small craft can enjoy the use of most, because sandbars nearly have fenced off their openings to the Sound.

We can trace most of them far northward across the plain. Bowman's Pond (Falmouth's Inner Harbor) lies in a channel which holds also Nye, Morse, Jones and Sols Ponds near Teaticket and which reaches as far north as Falmouth's Long Pond, which actually lies partly in the hilly moraine. Alongside the Inner Harbor to its east is Little Pond, the channel of which goes north to Jenkins Pond near the Sandwich Road. Great Pond ebbs and flows with the tides of the Sound some four miles. It lies in a valley reaching Camp Edwards and holds Coonamessett Pond and a long string of cranberry bogs in its northern part. Green Pond's channel goes nearly as far north. The eastern prong of Eel Pond at Menauhant is used as a valley by Childs River and reaches north of Johns Pond near Otis Field.

Childs River is using a hand-me-down valley, as are all streams

Figure 22. *An ancient flood channel. This is Childs River, Cape Cod, one of the broad, shallow drainage creases in the outwash plain.* B. CHAMBERLAIN

which flow in these channels. Today's streams are too weak and small to have scooped out the valleys themselves. Nor is it likely that meltwater streams carved them; none actually reaches the northern moraine whence such streams would have issued while the plain was being built. But during that barren period after the ice had recently left the land, flash floods from a moisture-laden atmosphere doubtless channeled their way across the surface. It has been suggested that such were the carvers of the present valleys.

Most of the small streams which now cross the plain flowing southward come from ponds to the north: Mashpee River from Mashpee Pond; Cotuit River from Santuit Pond; Coonamessett

River from Coonamessett Pond. Bass River, thought by some to have carried the Vikings into the Cape interior, rises in Follins Pond, less than three miles from the north shore.[9] A series of connected kettle holes takes Bass River to South Dennis, where it swings southwest to occupy another old stream channel. This is an extra large valley. It may well be a onetime spillway for the merging waters of the wasting ice.

Wind-Fashioned Rocks

An interesting pastime anywhere on the Cape and Islands is rock hunting; let us try it across this plain. As we stroll over the surface or dig through the soil, we may come upon small, polished, angular rocks. "Indian arrowheads," is our first conclusion, but on careful examination we see that these rocks are more roughly fashioned than a true arrowhead. Yet their surfaces are shiny and polished, as if buffed to perfection by a zealous bootblack; as smooth and greasy as talcum powder. Sharp edges separate the angular, flat facets. Etchings and flutings groove some of the surfaces, and each stone seems spattered with tiny pits.

Such rocks are called ventifacts (wind-fashioned). Once we are familiar with their features we notice them among larger rocks and, indeed, find them common. In fact, as physiographer William Davis pointed out in 1893: "in some cases they are so numerous and so large that they may be seen in the gravel banks from the passing trains; it being understood that the trains on the Cape do not travel at lightning speed."

Although related by birth to the outwash plain and common on it, ventifacts are not confined to this region, but occur in and on moraines and tills as well. From these environments they naturally spill out of sea cliffs to the sand of the beaches. Falmouth Heights has many, and they are common on the beaches of Nobska Point, Woods Hole and Sagamore Heights as well.

Ventifacts can take shape under any desert-like conditions where rocks are exposed to biting, wind-driven sands. On Cape Cod, most ventifacts are relics of those unsettled times when vast glaciers

[9] The interested reader is referred to the convincing evidence set forth by F. Pohle (*see* Bibliography II).

still stretched north of the present land and huge blocks of ice still rested upon it. Cold winds whipped around the flooding front of the ice and snatched up loose outwash sands into sweeping clouds of abrasive. These innumerable sands drilled, pitted and polished the larger rocks which they hit. Long and constant was the sanding. Gradually the peculiar angularity took shape, outlined by delicate edges and flat facets, like the unfinished efforts of gem cutters. The artisans' tools have long since vanished, but their work remains.

Curiosity prompts us to try to find these tools. We must look for a vast amount of loose sand that was active enough to leave its mark on solid rock. We should recognize such sand easily. Its grains, like the larger rocks it abraded, would have become etched and frosted. So we start our search with the Upper Cape land surface. There are no such accumulations of sand, except for that in obviously recent dunes. We try digging through the till and outwash, but again no luck. We even search in the floors of kettle lakes but do not find it.

So the mystery remains unsolved and the blasting sands remain hidden. We must guess that they are now on the sea floor, which was land during the late Pleistocene. It is interesting, though not conclusive, that scientists have found the right sort of etched and frosted sand blanketing a part of the bottom of Vineyard Sound.

Along exposed sandy parts of the coast, some sandblasting still occurs. Dry southwest winds prevail in summer and buffet the Cape and Islands. They sweep up the sands and hammer them against the rocks on the ground. A small number of ventifacts have formed and are forming this way. Winds which come from directions other than southwest are less effective, for they pick up moisture from the waters they cross, and this softens their bite.

Incidentally, the fruitless search for blasting sands in the floors of kettle holes is proof that the sandblasting occurred while the outwash plain was being built. When the fierce glacial winds ceased, a heavy blanket of the abundant windblown sand finally must have come to rest on the Pleistocene land surface. Some surely would have landed in the kettle holes if these had been open at that time as they are today. But as we have seen, such sand is absent there. We know therefore that when the abrasive settled to the ground, the ice blocks, like corks, still must have filled the holes; thus, the outwash plain still was being built.

Cape Cod's Moraines

Some time during the final stages of all this windblasting and kettle hole formation, ice surged again. This short-traveled thrust was its last. The three lobes gathered enough energy to push down from where they had marked time north of the Cape during the outwash plain building. They did not make much headway. They reached the northern and western edges of the plain. There they piled their loads, covering both gravel and, in places, still-unmelted blocks of ice.

Their recessional moraines are long piles of debris. Running like giant molehills parallel to the terminal moraine on the Islands, these Cape hill belts are similarly built of debris from mainland and sea floor. Thick and impressively massive, these moraines indicate that New England's mainland paid well with its substance for these ramparts of earth which now help to shield it from the attack of the open sea.

The trends of Cape Cod's three moraines, of course, tell us the positions of the three ice lobes which built them. During this advance of the scenery-maker, the lobes partly retraced their earlier advance to the Islands.

As we have seen, the Buzzards Bay Lobe was the most westerly. In its Cape sojourn it lay with its seaward edge along the Cape's Buzzards Bay coast. It reached across what is now the Cape Cod Canal to the mainland. Along this zone it built up the Buzzards Bay Moraine (*Figure 20*). This long ridge reaches from Woods Hole northward to slightly beyond the Canal; it is more or less parallel to the same lobe's terminal moraine, which makes up the Vineyard's western hills.

Along the present coast of the Massachusetts mainland, the Buzzards Bay Lobe adjoined the Cape Cod Bay Lobe. Together, along this joint, they piled up debris. This created the hills of Plymouth, such as the Pilgrims' Burial Hill, and such high hills as 395-foot Manomet Hill in the Pine Hills, on the western shore of Cape Cod Bay. This Plymouth interlobate moraine contributed Plymouth Rock to history.

Somewhere near the present Canal these lobes split. The Buzzards Bay Lobe went on to build the Buzzards Bay Moraine. The Cape

Cod Bay Lobe created the Sandwich Moraine along its southern front edge. This moraine runs parallel to the Upper Cape's Cape Cod Bay shore. It starts at Sagamore, where it mingles with the Buzzards Bay Moraine, and goes eastward to Orleans where it disappears altogether as the Cape bends northward (*Figure 20*).

As the Cape Cod Bay Lobe was joined on its west by the Buzzards Bay Lobe, so it was joined on its east by the offshore South Channel Lobe. The line where these last merged lay a short distance east of the present Outer Beach. Along this axis the two lobes shed and mingled their debris and meltwaters to build up the Cape forearm as far north as North Truro. Their deposits make up the interlobate moraine (*Figure 20*). The front edge of the South Channel Lobe stretched southward and, as we have seen, probably built up most of the ice deposits on Georges Bank.

The Recessional Moraine

The Sandwich and Buzzards Bay moraines are only parts of the ice sheet's recessional moraine. Paralleling the terminal moraine as it crosses Block Island to central Long Island, the recessional moraine snakes westward from Woods Hole, across the Elizabeth Islands to Long Island (*Figure 12*). Here a belt of hills, the Harbor Hill Moraine, runs along the north shore to Orient Point.

Cape Cod's Buzzards Bay Moraine

Approaching the Cape bridges, let us travel eastward on Route 6 along the northern or mainland side of the Cape Cod Canal. We see the skyline of Cape Cod across the water. A series of low-rolling, pine-covered hills and pockets, it looks just like the landscape along the mainland side of the Canal. Moraines lie on both sides, in fact, but they are different moraines. The Canal is a break between the Cape's Buzzards Bay Moraine and the Plymouth interlobate Moraine.

Cape Cod generally means Barnstable County, which includes the entire Cape and jumps the Canal to include the small area north of it as far as the Sagamore Highlands and the southern tip of Great Herring Pond. As we have seen, the glacial terrain which creates the scenery of western Cape Cod also continues a distance north of the Canal.

The Cape Cod Canal

The completion of the Canal in 1914 cut off most of Barnstable County from the mainland. And it provided such a clear-cut boundary to the Cape that forevermore, in the popular mind, you do not say you are on the Cape until you cross one of the bridges. Indeed, you may be told that the very air is sharper, clearer and more exhilarating on the Cape side. Perhaps it is. With the coming of the waterway, Cape Cod became an island and took for itself some of an island's pleasant climate characteristics. Older natives will tell you that the Canal has tempered the weather, winter and summer. It even has been claimed that on the southern side of the Cape one can bathe from March to December in the quiet bays! At any rate, the temperatures of the Cape and Islands are somewhat cooler in summer and warmer in winter than those of the mainland; and the Cape-Island growing season is nearly two months longer than that of nearby Concord.

Let us stop at one of the Route 6 highway turnouts to gaze at the placid, sun-flecked waters of the Canal. What a world of difference there is between this scene and that of eleven thousand years ago! To our north at that distant time, instead of Cape Cod Bay there was a huge, stagnating mass of ice. The Buzzards Bay Lobe probably had withdrawn almost completely from the region before the Cape Cod Bay Lobe retreated. As it too finally moved away, meltwaters poured from this lobe. Ice to the north and moraine piles to the east, south and west allowed the waters little chance of escape. Flooding the low area in front of the ice, they formed a now-extinct glacial lake.

The withdrawal of the Buzzards Bay Lobe had uncovered a natural break between the two moraines, a low area where they joined. The lake's waters rose until they found outlets. Some may have found their way through the moraine and flowed to the Sound via what is now the Bass River channel. A large amount of water, however, found the low area. This water flowed where the Canal now passes, across the narrow corner of the Cape into Buzzards Bay. A turbulent expanse of icy waters, muddy and laden with earth debris and occasional drifting rafts of dirty ice—this would have been our view here at that time.

Later, when the ice had vanished and the earth relaxed into a more pleasant era, two streams appropriated the old glacial outlet. Scusset River flowed northeast into Cape Cod Bay; and Manamet (or Monument) River flowed southwest into Buzzards Bay. Nowhere were these valleys more than thirty feet above sea level. A miniature drainage divide, a sandy ridge, separated them and kept their headwaters less than a mile apart.

The site of these two valleys was to become a logical one for a canal. The glacier already had done much of the digging work. And it was happily situated as well. By sailing through it, a vessel can avoid some sixty-five to a hundred miles of exposed and treacherous Cape backshore and shoal-filled sounds. The Canal, although privately built by August Belmont, is maintained today by the U. S. Army Corps of Engineers. The world's widest artificial waterway (540 feet), it forms a seventeen-mile connecting link between Buzzards and Cape Cod Bays.

Landscapes of the Buzzards Bay Moraine

Let us now cross the Canal on the Bourne Bridge. Thus arriving near the western end of the Cape, we may head south along Route 28. Soon we find ourselves climbing upward from the Canal lowland. We are climbing the ice contact slope, against which the front of the Buzzards Bay Lobe lingered awhile during its retreat. Before long we are on a hilly, irregular terrain along which the road runs for a few miles. We cross wide belts of till, patches of the ground moraine torn off from beneath the ice as it moved (*Figure 23*).

Surrounding the till is an irregular surface of outwash gravel. This, unlike the till, came not from the Buzzards Bay Lobe but from the Cape Cod Bay Lobe. For after the Buzzards Bay Lobe had built its moraine and begun its retreat, the Cape Cod Bay Lobe still was active. The meltwaters which issued from its western part were channeled into a narrow zone between the Buzzards Bay Lobe and the Buzzards Bay Moraine, and there they dumped their gravels. Fragments of rocks such as black and white slate, commonly found elsewhere in the Sandwich Moraine, never in the Buzzards Bay Moraine, are part of these gravels.

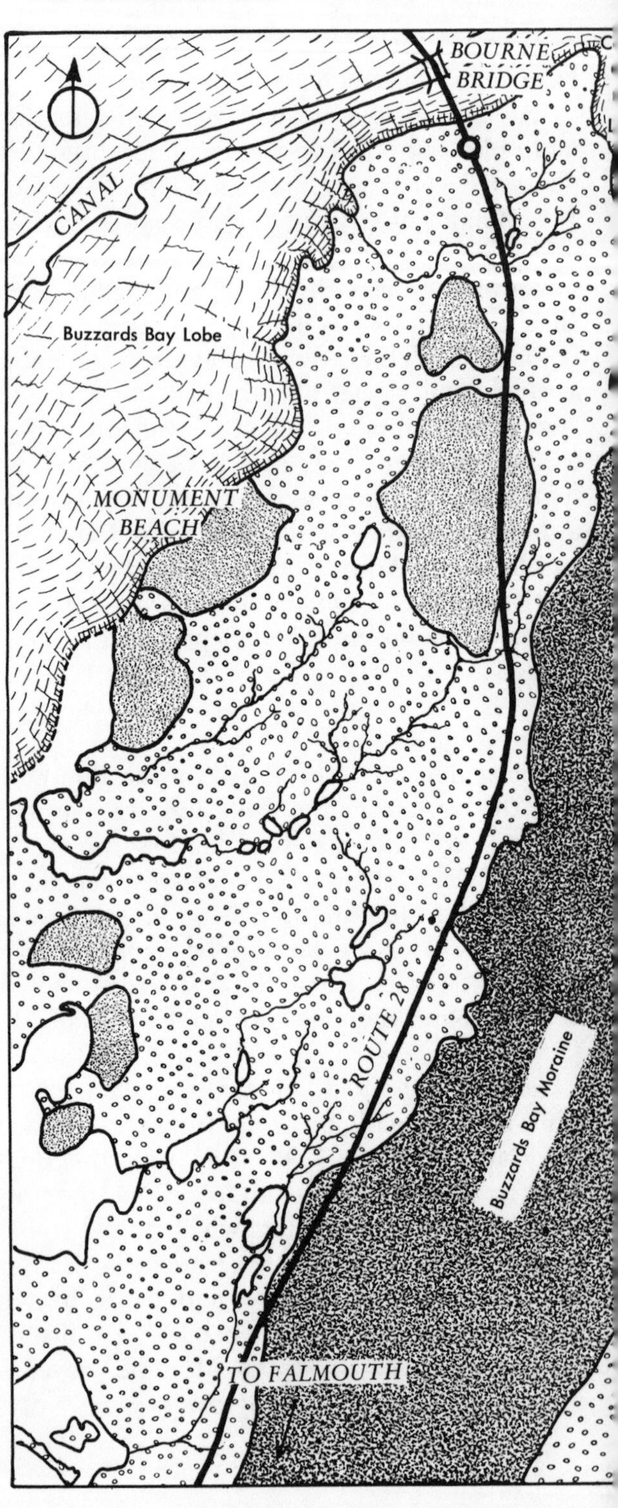

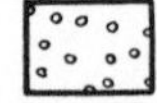

Figure 23. *Some glacial features of western Cape Cod. This sketch map shows part of the region of the Buzzards Bay Moraine. Between the moraine and the retreating Buzzards Bay Lobe spread outwash gravels from the still active Cape Cod Bay Lobe. Here and there rise patches of till left earlier by the Buzzards Bay Lobe.*

AFTER KIRTLEY F. MATHER, GEOLOGICAL SOCIETY OF AMERICA

As we cross this zone, the trees of the roadside veil distant views, but occasionally we may get glimpses of a ridge on our left, nearly parallel to but set back from the highway. It rises fifty to a hundred feet above the road. The ridge is the Buzzards Bay Moraine, Cape Cod's belt of western hills.

In every direction its topography bespeaks glaciation. Hills and hollows climb and dip without any predictable pattern. A couple of miles from the bridge rises the highest point of the moraine, Pine Hill (or Signal Hill) in Bourne, 306 feet above sea level or about a hundred feet above the highway at this point. As we go farther south the moraine drops in total altitude, although the bumpiness, the relief between hilltops and depressions, stays the same. Soon moraine and road approach each other and we find ourselves traveling directly along the western foot of the hills.

All along, between us and the shore, is ground moraine. Built on its drift are the shore resorts, fishing towns and old homes of Cataumet, Gray Gables, Monument Beach, Pocasset, Megansett, North Falmouth and West Falmouth. Harbors such as Phinney's, Red Brook, Squeteaque and West Falmouth Harbors, and most of the ponds in the region, are irregularities in the drift surface dating back to the time of deposition. Where such glacial depressions dip beneath Buzzards Bay today, they have been filled by fingers of the sea, risen since glacial times.

We climb with Route 28 onto the Buzzards Bay Moraine itself, shortly after West Falmouth. For about a mile we are on it. Then we drop down on its other side to the outwash plain and the town of Falmouth, which lies mainly on this flat surface. Long Pond, Falmouth's reservoir, reaches from the outwash plain into the moraine. We can picture the pond bed once filled with a long block of ice, which broke off from the glacier as it edged northward. This ice chunk must have remained throughout the chilly years during which the glacier lay north of the Cape, building up an outwash plain; and it must have lingered still during the period when the ice readvanced and left the Buzzards Bay Moraine surrounding it.

West Falmouth marks the general end of ground moraine. South of the town, this low, uneven surface is wedged out of the picture by a coastal indentation between Chappaquoit Point near West Falmouth, and Hamlin Point, north of Sippowisset. From Hamlin Point southward, the hilly Buzzards Bay Moraine itself lines the shore—in "masses of earth, sand, gravel and boulders tumbled together in the greatest confusion"—all the way to its final knobby protuberance at Woods Hole.

Let us follow the Buzzards Bay Moraine to its very end. We can pick up the Woods Hole Road in Falmouth. This will lead us back up onto the moraine, dip us over its knobs and basins, and land us at the head of Great Harbor, Woods Hole. Here the moraine leaves us shorebound while it hops across the water to become the Elizabeth Islands.

The harbors which nose into the moraine in this corner of the Cape are, like those in the ground moraine, flooded irregularities in its surface. The town of Woods Hole is named after the glacial lowland there which is now a tidal passage ("hole") separating Cape Cod from the Elizabeth Islands.

Rocks of the Buzzards Bay Moraine

Beneath the forested surface of the Buzzards Bay Moraine there are rocks, countless and varied. South of West Falmouth, the line of low bluffs and cliffs at the edge of the sea provides a cross section of the till and its boulders. Even inland to the north there are some patches where the forest has been stripped away and clusters of boulders laid bare. These are locally known as "bear dens" or "devil dens."

To a rock hound the southwestern end of the Buzzards Bay Moraine is particularly interesting. Here, unique among the Cape-Island ice deposits, is a peculiar type of granite. Generally weathered to a tan color, rather blocky, dull feldspar surrounds long lens-shaped crystals of gray, glassy quartz. The quartz crystals lie lined up in parallel rows and L-shaped figures. Like a rough Rosetta stone, the rock appears covered with ancient writing. Thus, it is called graphic granite. In forest rock piles and on beaches at this end of the moraine, this granite is particularly interesting because tiny red gems, clusters of small garnets, line its surfaces.

Interesting too are the other types of rock. The sea cliffs, especially at Woods Hole's Nobska Point, provide the widest assortment. The beach is littered with fallen giants. Long exposure to wind, sand and waves has given to many of them a dull, potato-like surface appearance. However, a simple hammer and chisel are, in many cases, enough to probe their interiors and bring to light their inner complexity and beauty.

The cushioning sand beneath these massive boulders foretells their fate. For, at one time, each grain of sand was itself an intrinsic part of some such crystalline rock. So, in time, will every exposed boulder of the moraine crumble before the forces of weather and erosion as it has begun to do this very hour, its rigid, colorful patterns collapsing into the relaxed homogeneity of sand.

Cape Cod's Sandwich Moraine

Natives call the Sandwich Moraine, the east-west moraine of the Upper Cape, the backbone of the Cape. As a backbone, however, it is made of soft cartilage rather like the backbones of sharks and skates than of hard bone like the backbones of most common fishes. Composed as it is of from five to more than thirty feet of loose till above soft outwash gravels, there is nothing rigid about the backbone of the Cape.

The Sandwich Moraine dominates the landscape as we cross the Canal, this time on the Sagamore Bridge, and travel east (*Figure 24*). Most of the Mid-Cape Highway (Route 6) runs along this moraine as far as Orleans; the Shore Road (Route 6A), on the other hand, runs mainly on ground moraine alongside the ridge and nearly parallel to it.

The Sandwich Moraine from the Mid-Cape Highway

We gain the best feeling for the massive bulk of the Sandwich Moraine by traveling on Route 6, along uninhabited, pine-covered miles. Much of this Mid-Cape Highway runs along the crest of the moraine, providing some magnificent views on either side and, on a clear day, a good panorama of the Cape from shore to shore.

Let us travel east, then, from the Sagamore Bridge. To the right as we leave the bridge, the land at first continues almost level with the road; the crowns of the trees which make up the skyline stretch nearly evenly southward, and we have no overlook. That high ground to our south is the Buzzards Bay Moraine. It meets the Sandwich Moraine at Bourne in a near right angle. Let us continue

Figure 24. *A typical morainal landscape on Cape Cod. Glacial drift provides abundant stones for stone walls.* B. CHAMBERLAIN

eastward a few miles, until we shall have crossed the width of the Buzzards Bay Moraine.

Now the views begin. We shall be overlooking the vast, pitted outwash plain for many miles. Where the moraine's contours fall and the highway drops in elevation, the roadside trees hide distant topography; but where the road climbs on the moraine's surface to above the heights of the treetops on the outwash plain, we see the irregular southern face of the moraine. It drops, appearing to flow like drapery, down to the flat plain over which we can look as far as the Sound if the day is clear.

The hummocky summits of the moraine rise from more than two hundred to nearly three hundred feet high between Sandwich and West Barnstable. Of these, the highest is 292-foot Telegraph Hill, easily spotted because of its fire tower. It rises about a mile and three-quarters southwest of the village of Sandwich. A climb up this tower provides a fine view of the moraine.

Southeast of Telegraph Hill rises Discovery Hill, some 260 feet high. From here, morainal knobs continue eastward but for a few miles diminish somewhat in height. At West Barnstable the highest hills barely reach two hundred feet. Then in Barnstable, the moraine puffs itself up to its most impressive dimensions and the Mid-Cape Highway makes its greatest ascent.

This hill is known as Shoot Flying Hill because from here gunners ambushed the great flocks of migratory birds. Now there are stopping places for cars beside the road and a small picnic area at the summit. We should pause here before following the moraine farther.

From Shoot Flying Hill, the Cape's unique sea-spiced beauty lies spread before us, in ice-sculptured relief far more impressive than any textbook illustration of glaciation (*Figure 25*). The pitted outwash plain below us to our south sparkles with kettle ponds. Lovely Wequasset (or Chequasset) Lake, the largest of these, lies spread out at the foot of the moraine. Now look west: along the distant edge of the outwash plain, running south, is a ridge with a rather even skyline; this is the Buzzards Bay Moraine. Let us turn north toward Cape Cod Bay: there before us, the Sandwich Moraine on which we are standing tumbles as haphazard hills toward the water. Beyond its northern edge, the towns which contribute their church spires to the view lie on the lower ground moraine, with its patchwork of salt marsh and tidal creek, all fringed by the white dunes of the shore.

Large boulders scattered in the picnic park on Shoot Flying Hill are typical of the coarser stuff of which the Sandwich Moraine is made. If we stand on the largest rock and look northwest, beyond the fire tower on the nearby hill, beyond the dunes and the Bay, we shall be gazing toward this boulder's birthplace, located somewhere beyond the farthest horizon our eyes can see.

Now we continue eastward on Route 6. The moraine drops noticeably in altitude. The highway takes us across the lower hills of Yarmouth, where a fifty-foot summit seems high, to Dennis, where we gain altitude again. Through Brewster and to Orleans the moraine becomes gentle, low and rolling, and we pass many kettle ponds (*Figure 37*) and dry kettle holes. When we cross Route 6A

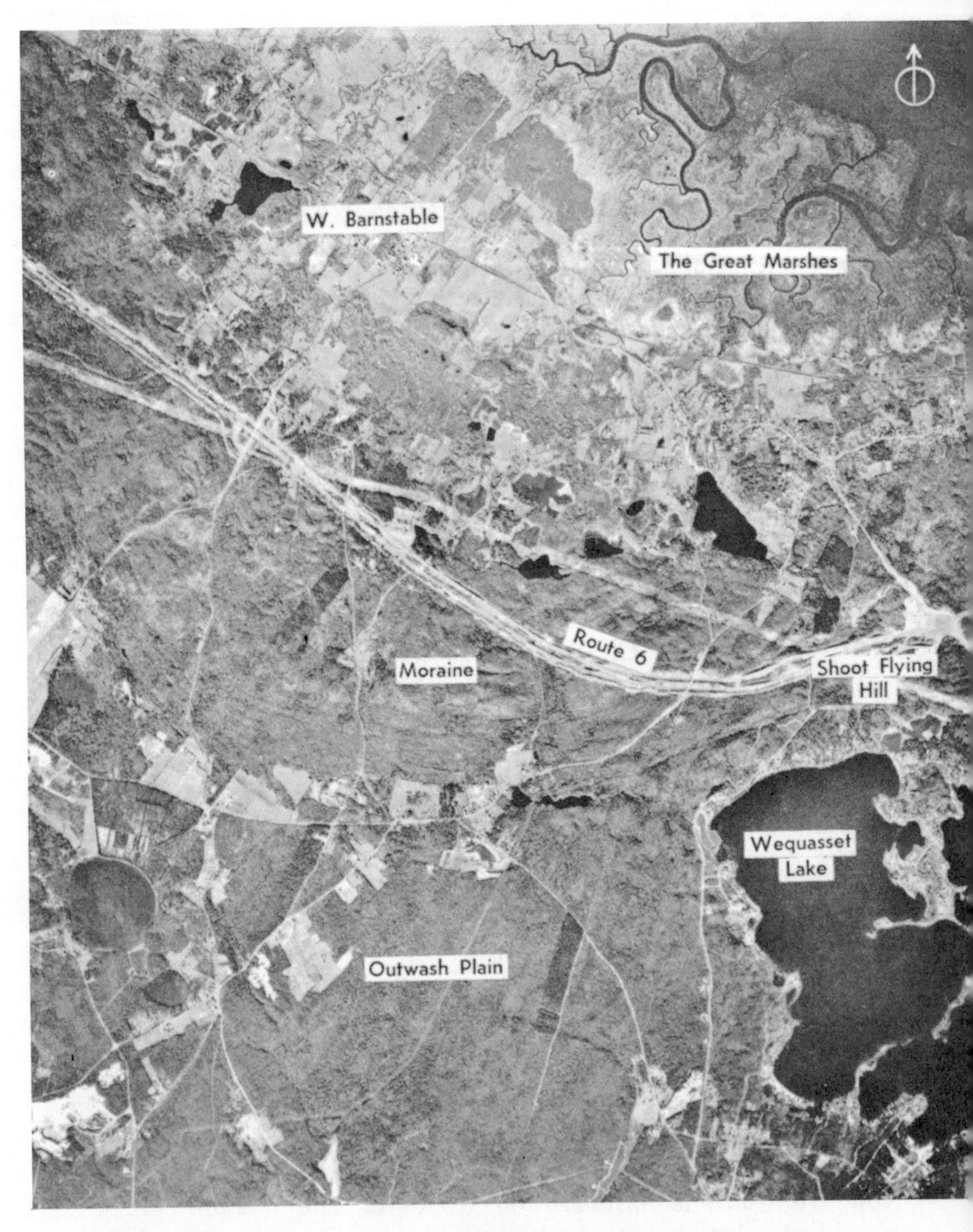

Figure 25. *A bird's-eye view of the Sandwich Moraine. This view over Shoot Flying Hill, Cape Cod, shows the moraine summits in the center, outwash plain to the south, and ground moraine, broken by marshes, to the north.*

U. S. COAST AND GEODETIC SURVEY

just before Orleans, we have reached the dwindling termination of the Sandwich Moraine.

The Sandwich Moraine from the Old Kings Highway

For a less lofty and more historical view of the moraine, we can go back and trace its length again, but this time via Route 6A instead of the Mid-Cape Highway. This shore road, the old Kings Highway, runs mainly on ground moraine. It takes us alongside the northern edge of the ridge rather than on it.

As we go eastward from Bourne, the Sandwich Moraine rises on our right nearly parallel to the road. To our left lie green expanses of marshland, fed by the tides. These represent low areas in the irregular surface of the ground moraine. Because of such widespread tidal drowning, this ground moraine appears neither as wide nor as well defined as that adjoining the Buzzards Bay Moraine. The cores of sandy-surfaced Town Neck and Scorton Neck in Sandwich and of Nobscusset Point and Quivett Neck in Dennis probably are ground moraine.

Along the western part of the shore road, the moraine's high hills may be impossible to see in summer due to nearby vegetation. Near Barnstable, however, we begin to catch glimpses of the ridge on our right as we go east. The town of Barnstable itself actually lies on the northern edge of the moraine. A few outlying morainal knobs rise here and there in town. There is, as in nearly every town along the Sandwich Moraine, a packet pole hill. Barnstable's was called Barrel Hill because of the onetime custom of announcing the arrival of the Boston packet ship by the raising of an empty barrel on the flagpole on this hill. As also in nearly every Bay shore town, there is a windmill hill, Cobb's Hill, on which a mill went up as early as 1687 to harness the sea winds.

Leaving Barnstable, we approach Yarmouthport and Yarmouth, where a series of low, rounded hills ripples the skyline to the south of the road. These are most visible in the revealing barrenness of winter. At Dennis, where the moraine again thickens, 159-foot Black Ball Hill and 160-foot Scargo Hill come into view. From the old highway, Scargo presents a stark, isolated silhouette against the sky. It rises behind Scargo Lake as if it actually were the great pile of sand which, legend claims, a tribe of Indians scooped from the lake bed to please an unhappy child.[10]

Scargo Hill deserves a stop for a good view of the moraine in this region. The stones which form the observation tower on the summit and the boulders that outline the parking area reflect the wide variety and large number of rocks to be found in the moraine. It is said that on a very clear day there is an eighty-mile view from the tower, although the usual low horizon haze cuts this about in half. Westward, hills peep over hills until the irregular, forested moraine merges with the horizon. Slightly west of south, the twin towers near Camp Edwards mark the Buzzards Bay Moraine. Southward, a level sea of treetops indicates the outwash plain. This begins a couple of miles south of the tower and reaches to the Sound. East and northeast, the moraine drops down again and levels off slightly to become a belt of small ridges, humps and dips, descending to meet the ground moraine and the Bay.

During the Revolutionary War, Scargo was one of the summits which served as relay points for messages between Signal Hill in Bourne and Plymouth across the Bay. The night skies over these hills often flickered with lights of bonfires as messages about British ship movements were transmitted, hill to hill and moraine to moraine, to Boston.

Eastward from Scargo, still following the northern edge of the moraine, Route 6A gently roller-coasters over its knobs or swings in wide arcs along its contours. The road passes through Brewster, which has settled itself near shore on the low, rolling ground moraine. West Brewster, however, is built right on the moraine and is thus bypassed by the shore road. To see it best, we must leave Route 6A and go to old Stony Brook Mill via Setucket Road.

Stony Brook flows from a kettle lake, the Lower Mill Pond, set in the moraine. This in turn is fed by a stream from an Upper Mill Pond farther south. Through postglacial centuries the little brook, flowing rapidly down the northern face of the moraine, has carved for itself a deep valley, so that the highlands of West Brewster rise abruptly on either side of it. For nearly three hundred years the mill has been grinding corn, using the brook to turn its waterwheel.

[10] Scargo, the daughter of a chief, had some pet fish which she kept in a small pond. When the pond dried up she was so upset that the tribe dug for her a new one, its length and breadth determined by the flights of arrows. Scargo Hill is the debris dug from the lake site.

A series of man-made dams controls its flow. Below the mill, north of the road, Stony Brook wanders through marshlands to Cape Cod Bay.

The footpaths which wind along the brook south of the mill go past some of the largest erratics visible in the entire moraine. Great boulders—immigrants, apparently, from northern Massachusetts and New Hampshire—appear on both sides of the stream, but the largest is on the eastern shore. Slowly, bit by bit, the flowing water has removed the clay and sand of the till which once surrounded these rocks, and has left them jutting out (*Figure 26*). Were other streams to eat their way through other chance parts of the Sandwich Moraine, countless similar huge erratics undoubtedly would come to light. We can only speculate how large the largest might

Figure 26. *Stony Brook, Brewster, Cape Cod. This valley was cut into the Sandwich Moraine, leaving glacial erratics exposed along the shores.*

B. CHAMBERLAIN

be. The boulders along Stony Brook so impressed the Indian tribe which lived, under the leadership of the sachem Wono, along the shore of the brook,[11] that they put the largest rock to work as a prayer rock and conducted tribal ceremonies around it.

Back to Route 6A now, we find the moraine bordering it from Brewster to Orleans low and subdued. The lakes and boulders of Nickerson State Park are parts of it, of course. From the fire tower in the park we can appreciate how the gentle topography here differs from those parts of the moraine visible from Scargo and Shoot Flying Hills.

A slight surface irregularity at Orleans and an abundance of boulders in the fields are the only evidences of moraine there. The uneven border of the drift in that region creates the pockets and grooves in the shorelines of Town Cove, Little Pleasant Bay and Nauset Harbor. Town Cove itself may be a submerged chain of kettle holes similar to such chains in the outwash plain. Little Pleasant Bay and Nauset Harbor are protected on their east by sea-built sandbars; the moraine itself ends in the rather abrupt bluffs which line the "necks," "points" and islands in this area, rising about sixty feet behind the sandbars.

We can follow the moraine to where it peters out. We travel along Beach Road from Orleans through East Orleans to Nauset Beach. In this region it has most definitely fallen from its former state and scarcely deserves the name of moraine any more. Near the beach the road descends from the ice deposits and crosses a narrow neck of the extensive marshes which, like the beach itself, came into being in postglacial times.

From the waters of the lovely salt inlet which reaches from Orleans to Chatham rise small morainal islands—Sampson's, Pocket, Hog and Sipson's—to "add beauty to the haven," as J. P. Freeman wrote in 1802, "and give it a just title to the name which it has received, that of Pleasant Bay."

At this end of the Sandwich Moraine, the front of the Cape Cod Bay Lobe apparently stretched farther south than elsewhere. A finger of moraine is traceable southward, creating the landscape of Chatham as far as the Sound. Some of the beautiful panoramic views to be had in Chatham itself are from morainal hills.

Abundant glacial boulders lie sprinkled over the morainal land near and in Chatham. Low drift creates the ups and downs of

[11] To the Indians it was Sauquatucket River.

Chatham Town. These are bordered on the west and north by the knob and kettle topography culminating in 131-foot Great Hill next to Lovers Lake.

North of Orleans, the Sandwich Moraine yields up its individuality and merges almost imperceptibly with the sand plains of the narrow Lower Cape.

Cape Cod's Interlobate Moraine

From northern Orleans to North Truro stretch these windswept plains of the Cape's north-south arm. They were formed of deposits left when the sides of two merged glacial lobes melted. Thus, most of this Lower Cape is interlobate moraine. This moraine, reaching from shore to shore and facing open ocean, is particularly vulnerable to winds and sea, the attackers of the land. So readily does it yield to the attack that today's Lower Cape is, on a geological time scale, merely a passing frame in a colossal, continuous motion-picture reel. Its map shape and surface features are momentary and transient, endlessly changing as the land evolves.

Along the shores, the sea has wrought impressive changes, even within the memories of inhabitants. The land itself has been forested and deforested, plowed and overplowed, has become desert and wasteland, and is reverting today to moor and forest again.

Early Appearance of the Lower Cape

When the earth warmed and the ice withdrew from the land to feed a rising sea, the Lower Cape's appearance was considerably different from its hooked shape of today. At first there were only ice deposits. The sea had not yet filigreed the shoreline with beaches, bars, spits and forelands, and there were no marshes or salt creeks. All of Provincetown and Monomoy were missing, for these are more recent sand deposits.

The glacial shorelines were highly irregular. Seawater crept in wherever the hummocky drift was thin and low. Before wave action could plane smooth much of the coast, its pattern was closely similar to the irregular coastal pattern which has remained undisturbed in peaceful, protected areas, such as quiet bays, today.

In addition, the Lower Cape was considerably wider at that time. The sea had not yet made inroads into it. In the subsequent process of narrowing, more drift has been eaten away from the ocean shore than from the bay; for the former has been exposed to the pounding action of the open sea. Cliffs such as those at High Head, North Truro, which now hide from the sea's attack behind postglacial sand deposits, must have been eaten back to their present positions before such sandy protection came about.

Using such criteria as these, William Morris Davis and Douglas Johnson, in their separate classical studies of the region's shoreline, reconstructed the original outlines of the Lower Cape as shown in *Figure 27*.

Figure 27. *The possible original shape of the Lower Cape. This shows two reconstructions of the Lower Cape immediately following glaciation. The dashed line is the outline suggested by William M. Davis; the solid line is that suggested by Douglas Johnson.*

DOUGLAS JOHNSON, THE NEW ENGLAND-ACADIAN SHORELINE

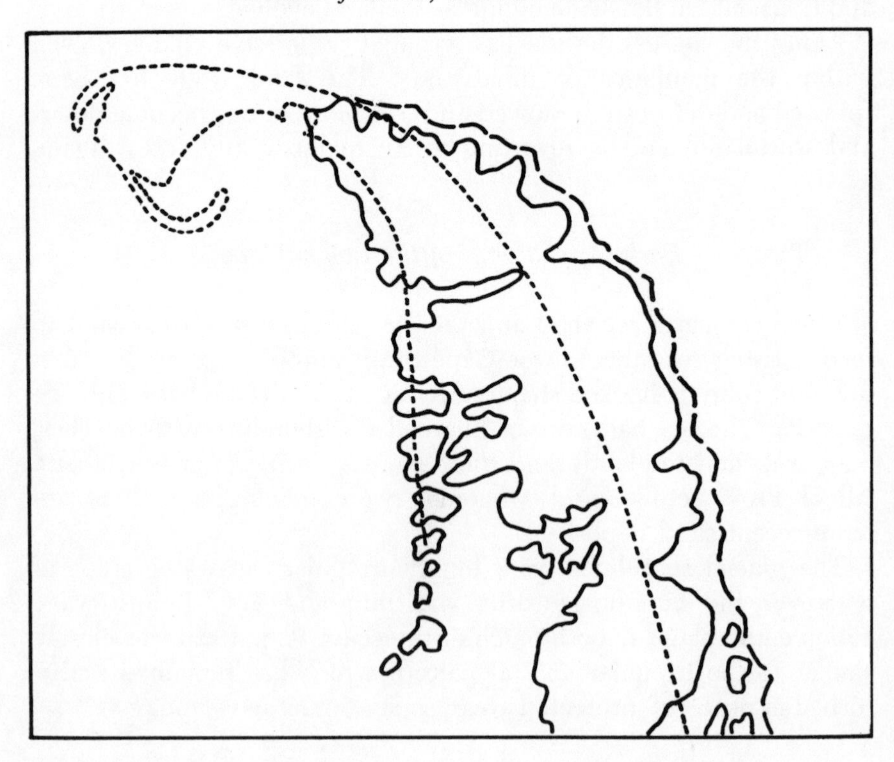

Sections of the Interlobate Moraine

The surface of the entire interlobate moraine is covered with ice deposits. These fall into three geographical divisions, based on elevation and appearance of the landscape (*Figure 20*). Across the central area stretch the High Plains of Wellfleet-Truro. Nearly along the southern boundary of these plains runs Lecounts Hollow Road in South Wellfleet, emerging at McGuire Beach just north of which the southern edge of the High Plains meets the sea.

The High Plains overlook to the south of McGuire Beach the nearly flat surface of the Plains of Eastham. These, in their southern part, give way to low hills and finally merge with the eastern edge of the Sandwich Moraine at Orleans.

North of McGuire Beach the High Plains reach as far northward as Cape Cod (or Highland) Light, North Truro, beyond which they drop off by a definite dip in the landscape to the lower North Truro Plains.

Two Possible Origins of the Landscape

Thus the Wellfleet-Truro High Plains are flanked by comparatively flat lower plains on either side. It seems as if these lower plains may have been eaten back into the High Plains by some sort of erosion following the last glaciation. Indeed, the individual beds of sandy drift at the *surface* of the lower plain can be traced laterally to where they continue into *subsurface* beds of the High Plains—as if the beds now above them in the High Plains have been peeled away from the lower plains on either side.

Geologists hold differing opinions concerning the erosion responsible for this peeling. One theory suggests it was the work of the sea, and occurred when the land was lower and the sea lay over it. It further suggests that wave action took place on all three topographic regions of the interlobate moraine. This requires two separate stages of such wave action and therefore two different times that the sea crawled over the land.

Blanketing the surface of the High Plains is the possible evidence for the first of these sea invasions. This consists of irregular, thin glacial drift lying upon a relatively flat surface. The irregularity and natural aspect of the drift indicate that it never has lain beneath seawaters, which would have rendered it unrecognizable. But the flat surface beneath it indicates that on that flat surface, at least, seawater did lie, smoothing and flattening it. Thus, there is indicated a sea invasion of these High Plains, followed by an ice invasion. The sea invasion required the present land to be nearly 140 feet lower (or the sea that much higher). Then the waters left and the ice came. It actually need not have covered this area to have left such deposits as are found; perhaps they originated in meltwaters or dropped from floating ice sheets. Later, the land must have been uplifted, not just 140 feet, but nearly 200 feet. The ice melted and the sea rose, leaving the lower fifty feet below sea level and more than 140 feet above it, as we have today.

Then, suggests this same theory, came the second sea invasion. This time, apparently, water did not cover the surface of the present High Plains, for they already stood well above that sea level. Rather, waves ate their way across the north and south sections of the High Plains, planing them down to the lower level of the North Truro and Eastham Plains of today. These lower plains are flatter, for the smoothing effect of water, rather than the rough, disorderly activity of ice, was the last process to mold their surfaces. The original drift appearance here has nearly been destroyed, and only some entrenched glacial features, such as outwash channels, remain.

Many geologists do not accept this wave theory. They point out that the westward slope of the plains' surfaces makes the theory untenable. Sea action should have caused the plains to slope gently seaward, as any offshore sea floor slopes, and this would be to the east, not to the west. They also consider excessive such an uplift of the land as nearly two hundred feet between ice retreat and the present time. A better explanation, say these geologists, is some sort of stream action. Streams could have created the same types of surface features. In fact, a resemblance has been noted between these plains on Cape Cod and the sandy plains of Iceland, which were simply formed on the land surface by meltwaters from ice.

Ice Debris on the Lower Cape

However the plains were formed, there is no question as to the glacial nature of their drift. During the final glaciation this drift came partly from the Cape Cod Bay Lobe and partly from the offshore South Channel Lobe which lay east of the present coast. Thin though this final drift is, it is interesting. Its motley pebbles provide not only the sole available clues to the path of movement of the South Channel Lobe, but also indirect evidence of the materials of the sea floor off northern Cape Cod.[12]

To look closely at a sample of this uppermost drift, we dig down from the surface to get it. But we shall have difficulty spotting where till ends and the intertill beneath it begins, for they are similar in appearance. As we have seen, a sequence of tills and intertills lies spread across the Lower Cape, deposited during the Manhasset glaciation. There is scarcely any visible difference between any of those three ice deposits (Tills 4, 3 and 2) and this most recent drift blanket (*Figure 10*). All originated in very nearly the same source areas.

However, within the most recent drift there is a zone of division splitting it into two parts. The lower has been called Till 1-a; the layer above it, Till 1. The distinction between these is more clear than that between the lower of these and Till 2 of the Manhasset stage beneath. Till 1-a is sand, gravel and pebbles, with almost no large boulders. Till 1 likewise consists of sand, gravel and pebbles, but has in addition a smattering of boulders and erratics. Since this bouldery layer is the uppermost of the two it was deposited last. Till 1-a is the remains of weathered rocks, soils and earlier glacial debris scraped from the mainland by ice at the beginning of its third and final major advance over this region.

It may well be that the ice was leaving lower Till 1-a strewn across parts of the Cape as ground moraine while it was busy building the terminal moraine on the Islands. Then, as we have seen, it retreated briefly some distance away from southeastern New England before gathering strength for its last advance, which led to Cape Cod's recessional moraines. On this final advance it crossed part of the mainland again—a mainland already swept nearly clean of loose debris. Grist for this ice had to come from larger rocks which lay exposed at the denuded surface of the land. These larger rocks became part of the Cape recessional moraines and part of Till 1, the uppermost till in the interlobate moraine.

When we examine particularly the rocks of which the pebbles in these upper tills are made, we see that they resemble those of the

[12] This is discussed in detail in Chapter VII.

nearby Sandwich moraine more closely than they do those of the Buzzards Bay Moraine (*Figure 34*). We could expect this, since the Cape Cod Bay Lobe, which built the Sandwich Moraine, also contributed part of the drift of the interlobate moraine. Among these pebbles, granites and other bedrock from the mainland mingle with fine-grained volcanics mainly from beneath present offshore waters. Volcanic rocks, in fact, make up nearly half of any sampling of stones. There is clay too, some of it much like the Gardiners Clay.

These pebbles and the clay indicate different sources for this moraine's till. For consider the paths and relationships of the lobes during their greatest activity. The edge of the South Channel Lobe lay somewhere east of the Cape's present eastern coast. It and the Cape Cod Bay Lobe, although joined as one ice body to the north, were making their separate ways in front by the time they reached the Cape Cod region. Near or along the present Cape forearm there was a long narrow crack between them. Each lobe, as it moved southward from the split, at the same time fanned out radially. Each lobe contributed part of its till to this interlobate zone.

Apparently neither had much to contribute. The ice did not have much power to erode along this zone, so that little till was acquired in the first place. Indeed, the debris left on the Lower Cape during the final glaciation is so thin as to be scarcely significant in places. The region owes its rolling uplands and sudden cliffs rather to the substantial foundation of Gardiners Clay and Manhasset tills and intertills.

The South Channel Lobe must have lain with its western edge not much farther than two miles beyond the present eastern Cape shore. We can estimate this by considering the erosion which has taken place since then. If we know the former extent of the land, we know the approximate location of the ice, since the ice deposits were built up near the ice border. The eastern coast from Chatham to Truro has borne the brunt of an unceasing attack by winds and waves these ten thousand years since the ice left. Originally, therefore, the deposits must have reached farther east. Along substantial sections of this coast, it has been estimated, erosion today proceeds at some three and a half feet each year. Thus, land once extended some two miles farther out, and that is where the ice must have been.

East of Cape Cod, the shape of the sea floor outlines the onetime South Channel Lobe for us, with its margin slightly beyond where open sea now washes the Cape forearm. A submerged sloping surface off the Cape's eastern coast rises like the rim of a platter toward the west and south, climbing away from the area once occupied by ice. Apparently this represents the contact border between ice and land when sea level was lower. The most seaward part of this ice contact slope is to be found on the north side of Georges Bank. It follows the edge of the Bank westward, then bends northward to swing up along the eastern coast of the Cape.

On Cape Cod at North Truro, the glacial deposits drop from sight. But they continue submerged still farther north. There is a hill covered with eighty feet of water north of Provincetown, known as Stellwagon Bank. It rises more than six hundred feet above the Bay floor and is covered by drift-like deposits. Its shape, trend and materials indicate that it may be a lower continuation of the same wedge of debris as that which forms most of the Lower Cape.

Rocks on the Lower Cape

Large rocks, although more abundant, as we have seen, in Till 1 than in lower Till 1-a, actually are not common to any part of the Lower Cape's drift. Accordingly, the few erratics which do exist here have been awarded a modest degree of fame. As you approach Cape Cod Light, North Truro, for example, you see a ten-foot boulder lying in quite lonely grandeur beside the road to your right. The cottage behind has long borne the name, "The Rock," although the words at this writing are weathered nearly beyond legibility. A boulder of this size is a most rare occurrence on the North Truro Plains.

But Eastham's Enos Rock makes up in size for the scarcity of rocks on the Lower Cape. Enos (or Enoch's) Rock is the largest erratic on the Cape—perhaps the largest in southern New England—and the town has set up a small park around it (*Figure 28*). The rock lies just to the right as one travels on Doane Road from Route 6 to Nauset's Coast Guard Beach. This huge boulder, forty-five by eighteen by twenty-five feet, is a block of crystallized volcanic lava, similar chemically to basalt, and weathered on its surface to a dull, greenish-gray color. Because of its size, it is unlikely, although not impossible, that this rock was carried the entire

Figure 28. *Enos Rock. This block of lava at Eastham, Cape Cod, is possibly the largest glacial erratic in southeastern New England.* B. CHAMBERLAIN

hundred miles from the mainland. More probably it was torn loose from the belt of volcanic rocks which seems to lie beneath part of Massachusetts Bay.

Occasionally other large boulders in the till become exposed in the cliffs by waves and eventually fall to the beaches of the interlobate moraine's shores. Along Nauset Beach from Eastham northward there are always a few large boulders a little below high-tide level, which appear and vanish like ancient wrecks at the whims of the shifting sands.

The Plains of Eastham: The Kame Belt

The Plains of Eastham (*Figure 20*) are mostly as flat as we should expect plains to be. But in their southern part, where we first meet them as we leave the Sandwich Moraine at Orleans, knobby hills and saucer-like depressions constitute the landscape. These are kames, distinctive sorts of glacial hills. They make up the rolling, cedar-covered fields, undulating grassy meadows and pine forests of the Eastham area. The town of Eastham itself sits amid such hillocks.

Although kames are confined to morainal belts, they consist of a sort of layered drift which is more orderly than true till. Let us consider conditions while the ice wasted back and poured its waters onto the land beyond. Streams issued from its towering front, of course, and snaked through subglacial channels. But others, likewise born of the warmth of the sun, flowed across the top of the ice. These, like all glacial streams, carried solid earth materials which had been imprisoned in the ice. As they reached the edge of the glacier and spilled over in waterfalls, they dropped their loads as conical piles of partly sorted drift, or kames.

Northward to just beyond Nauset Light the kame belt continues. Here it is sharply truncated by waves into massive, ever-changing ocean cliffs, and north of here the hills give way to the level Eastham plateau. The boundary between kames and plateau runs across the Cape on a southwesterly slant, from just north of Nauset Light to near Kingsbury Beach on the Bay. The Ocean View Drive from Nauset Light Beach to Coast Guard Beach, Eastham, takes us across nearly the entire width of the kame belt; its southern edge reaches the ocean at the former Coast Guard station. South of this, if we were to cross the marshes, we should find ourselves back at the Sandwich Moraine. The borderline between kames and moraine runs southwest also, across the Cape arm from the former Eastham Coast Guard station nearly to Rock Harbor, Orleans.

The presence of kames in the southern part of the interlobate moraine indicates that while they were forming, the ice lay stagnant and melting immediately adjacent to, not on, the kame belt. Yet several pond-filled kettles, such as Great, Herring, Depot, Mudd and Minister Ponds in Eastham, indicate that prior to the kames, ice must have covered the region.

Cutting through the kame belt along a cliff-edged lowland there is a reed-lined salt creek, Boatmeadow Creek, half a mile north of the Orleans-Eastham border. It winds across the Cape from near Town Cove to Cape Cod Bay, occupying a channel which both nature and man took part in digging.

It began with a natural break in the ice deposits, a low area due to happenstance deposition irregularities. Gosnold, in 1602, referred to the Lower Cape as an island, indicating that water bisected the Cape here. Even a century later, in 1717, a whaling boat was able to sail completely across the Cape along Boatmeadow Creek during a time of high water; and small boats carrying early colonists cut through the Cape here enroute from New England to Virginia (*Figure 29*).

Then in 1770 came a "furious gale of wind which was accompanied by a tidal wave" (wrote C. Townsend of the Coast Survey in 1890). "This herculean effort of the elements changed the whole east and south shore of the Cape, and deposited, in the salt marshes and lowlands, sand hills sixty feet high, and completely washed away a sand point off Nauset, where to this day [1890] at extreme low tides, stumps of former trees have been laid bare, which have been seen by men . . . who have visited the spot for that purpose." And boats no longer could use the Boatmeadow Creek shortcut.

But Cape Codders had wanted a dependable canal through their peninsula since the days of Washington. Indeed, they had toyed with the idea as early as 1697. Early in the nineteenth century they put shovels to the Boatmeadow Creek lowland, then filling with marsh plants. They widened the channel and deepened it; and in 1804 again gave the sea access.

That first Cape Cod Canal became known as Jeremiah's Gutter. From Eastham northward, Cape Cod became an island again. Jeremiah's Gutter had its day. Sailing ships could avoid the hazards of the trip around the very dangerous backshore of the Cape. During the War of 1812, Cape salt producers used the passage to evade the British blockade. But then a dike was put up west of the railroad track, the Canal became neglected, and marshes and sandbars closed it up and tied the Cape together again.

The Plains of Eastham: The Tablelands

From Orleans, let us go north to Eastham, where we leave the kame belt. The knobby topography gives way to a remarkably even,

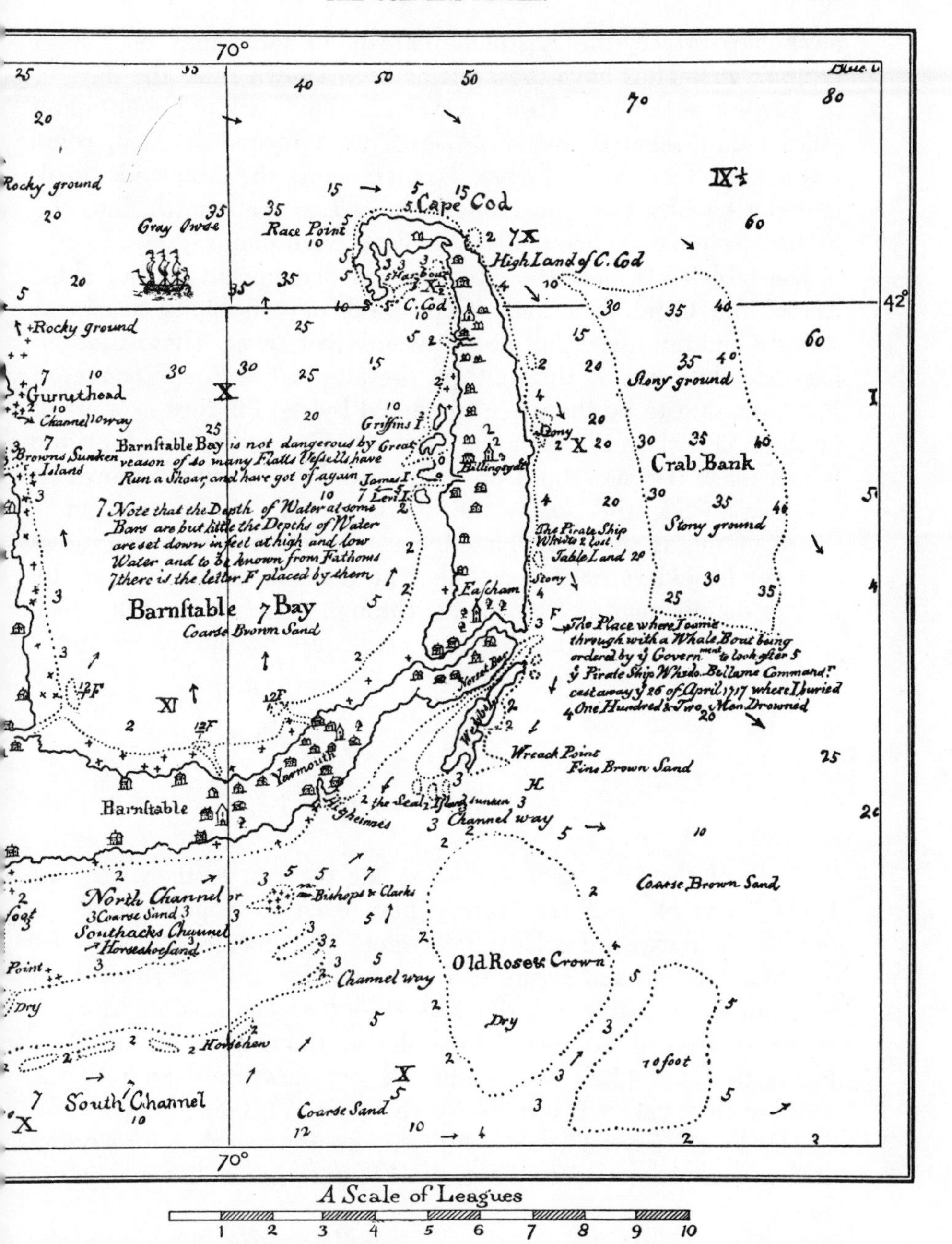

Figure 29. *When the Lower Cape was an island. This eighteenth-century chart of Cape Cod shows the Upper Cape separated from the Lower Cape by the Boatmeadow Creek cut-through.*

HARVARD MUSEUM OF COMPARATIVE ZOOLOGY

mesa-like surface, the Eastham plateau, or tablelands of Nauset (*Figure 20*). Thus have these plains been known since the days of the earliest settlement. Always a welcome landmark to homecoming sailors, they seem to hover, as level as a second horizon, some seventy-five feet above the sea. To their south, the tablelands break into the knobby Eastham kame belt; and to their north, into the higher, irregular skyline of the Wellfleet-Truro plains.

The tablelands meet the Atlantic in a magnificent wall of cliffs. Across the Cape they meet Cape Cod Bay in bluffs nearly as straight and unbroken, but about twenty feet lower. These straight, bay-side cliffs end on the south at the Site of the First Encounter, Eastham, said to be the place so named by the Pilgrims as a result of their first skirmish with the Indians. The boulder and plaque which mark the site stand on a forty-six-foot headland known as Mulford's Cliff; this forms the southern end of the tablelands. South of this is the irregular edge of the kame belt, swinging in and out in wide arcs of coastline. The lowland at the foot of the cliff is mainly marine marshland, through which lazes salt creek Herring River. These marshes and creeks are postglacial creations of currents, sand and vegetation.

The High Plains of Wellfleet-Truro

North of the tablelands and in strong contrast to them rise the High Plains of Wellfleet-Truro. They reach their greatest elevation on their seaward side. A mile south of Pamet River, they rise nearly a hundred and eighty feet above sea level, sloping westward to about a hundred feet at the Bay. Here they end in a labyrinth of necks, coves and marshes. Along the northern edge of the High Plains the land takes a one-hundred-foot downward step to the more even-surfaced Plains of North Truro. This escarpment runs southwesterly from Cape Cod Light, on the North Truro ocean cliff, to the Cape Cod Bay mouth of Truro's Pamet River (*Figure 20*).

Broken by kettle holes and wide, parallel valleys, the hummocky High Plains rise and fall like a brown and green turbulent sea (*Figure 30*). They are among the wildest, most picturesque parts of Cape Cod. Covered here by pine trees and there by a spongy

Figure 30. *The rolling High Plains of Wellfleet-Truro.*
THE NATIONAL PARK SERVICE

carpet of bearberry today, this region—like much of the interlobate moraine—was once fertile enough to farm.

The topography of the High Plains, unlike that of the Eastham and the North Truro Plains, is obviously morainal. Late Wisconsin-Stage drift, as we have seen, lies on the eroded surface of earlier tills and intertills.

To view the sea from the eastern edge of the High Plains is one of the thrilling experiences the Cape has to offer. Some maintain that nowhere is there a more spectacular or beautiful panorama of the Atlantic's waters. Certainly the changing moods and infinitely varied kaleidoscope of colors which drift across the face of the sea before these cliffs are magnificent. And nowhere is the view more impressive than in North Truro by Cape Cod Light. Here, at the very northern part of the Wellfleet-Truro High Plains, the banks rise 130 feet above the beach.

Up at the top of these banks, the most recent glacial drift forms the light-brown capping, a few feet thick. This supports the thin soil which hangs over the cliff edge, held in place by sheer root entanglement after waves have undermined its foundations. In the shadow of the overhang are wind-formed blowholes, like rows of fortress windows, where the bank swallows nest. Below this is gravel and Gardiners Clay, which, as we have seen, is the major foundation of the cliff here.

Let us wander southward along the beach from here. Soon we notice another feature of the High Plains. This is the series of valleys which cut east-west across the Cape forearm and, opening onto the beach, scallop the sea cliffs. In a shore otherwise lined with sheer, often insurmountable bluffs, they afford the only breaks. These have long been called "hollows." Thus, the word, hollow, has come to be applied to the valleys themselves (*Figure 31*).

If we follow the valley of one of these hollows inland, we find it to be very wide and deep. Many are followed by streams today, and in them the little brooks appear as misfitted as children dressed to play grown-up. Obviously none of the present streams are powerful enough to have carved such valleys.

But once a brook has found one of these channels it will not readily leave; for running water, for all its potential power, is lazy if given the chance to be, and will use any available channel for a bed rather than make its own.

Not all of the channels are occupied only by brooks; many roads which cross this part of the Cape also take the easy way and follow dry valleys. Route 6, on its way along the forearm of the Cape, dips in and out of such grooves, indicated by roads leading off the highway along them. Most of the hollows have names, and the larger are conspicuous enough in the Cape landscape to warrant listing.

The northernmost, Dyers Hollow, is a channel followed for some distance by Little Pamet River, then by Long Nook Road. About a mile and a half south of this is the valley of Pamet River. Pamet actually is an estuary of the sea, flowing with the tides. Just south of Pamet is Brush Valley, followed in part by Pamet Road South. Newcomb's Hollow belongs to a channel which runs east of Herring River in Wellfleet. About half a mile south of this is Pearce's Hollow. Farther south, Cahoon Hollow Road follows the valley, Cahoon Hollow, from Great Point eastward to the Cahoon Hollow Coast Guard station. Two miles south of this is Snow's, or Lecounts Hollow, the channel of which is followed by Blackfish Creek and Cook Road, and which reaches the sea at McGuire Beach. Still farther south, Fresh Brook follows the only important valley to break the surface of the lower Nauset tablelands in their eastern part.

Lying on the floors of some of these valleys, and surrounding all of them, is drift from the scenery-maker. Thus we know that they originated between the final glaciation and the preceding Manhasset. During that interstadial period, heavy streams ate their way into the soft glacial debris, carved the channels, and then dwindled in size or ceased to be. When the ice returned, some of its debris found its way into the then-untenanted channels and has remained in their lower parts.

Pamet River is a good example. In fact, it has given the term, pamet, to geology; a pamet has come to mean a channel in sand and gravel deposits, resembling a stream valley but with its sides and bottom covered with irregular glacial drift.

The gradients of these hollows run from east to west. Thus, we know that the waters which carved them flowed east to west. Since sea cliffs bevel the valleys today in their upper eastern reaches, the valleys once must have reached still farther east, somewhere beyond the present coast.

Probably the source of their original streams was the South Channel Lobe of the glacier on its Manhasset journey south. As we know, this lay east of the present shore, while the other lobes were building terminal moraines on the Islands. As it stagnated, this ice fed meltwater streams which flowed westward from its edge, carving deep channels in the previously leveled plains.

Sea level at that time was lower, of course, due to glaciation, and the streams channeled their way down to the base level of erosion provided by that sea level. Some of them reached or neared this base level, for the bottoms of some of the channels today lie below the present sea level.

Beyond the drift plains, the streams spilled into the broad lowland, now Cape Cod Bay, which already had been vacated by the Cape Cod Bay Lobe which had scooped it out. As the South Channel Lobe on the east also ebbed away, the powerful streams dwindled and died. Then the wavering ice readvanced and, as we have seen, covered the surface with ice blocks, piling drift everywhere.

Providential indeed appeared these hollows in the cliffs to shipwrecked sailors of past centuries who managed to reach the beach, worn and often injured (*Figure 31*). Directions for locating them were provided in Blunt's *American Coast Pilot,* standard equipment on early American ships; for the dangers of the Cape Cod coast had been fully realized and feared from the earliest days of American sailing. In the late eighteenth century the Massachusetts Humane Society, forerunner of the coast guard, built shelter huts stocked with fuel and food in seventeen of these hollows.

Crossing the western half of the plains are other valleys similar in origin. Some of these open into hollows in the bluffs along Cape Cod Bay. Paradise Hollow, Lombard's Hollow and Prince Valley are the most conspicuous.

Figure 31. *Where the Wellfleet-Truro High Plains meet the sea. This view shows one of the hollows which permit access from the beach to the surface of the plains.* NATIONAL PARK SERVICE

Along the Bay shore of the High Plains, the irregular outline of moraine has provided the Cape with one of its better harbors, that of Wellfleet. The sea has made it complete by throwing up a postglacial assortment of sandspits and sandbars. The knobs of marsh-enclosed land known as Bound Brook Island, Griffin Island, Great Island, and Great Beach Hill—those rounded hills one sees driving to Chequesset Neck, Wellfleet, or from the Wellfleet fire tower—all deserved the title of "island" originally. Piles of ice debris, separated by low areas, they became islands when the postglacial rising sea isolated them. Through the centuries they remained islands, until as recently as 1780. The sea is an active builder, however, and saltwater vegetation thrives in such shallow, protected waters as those between the hills. Today marine marshes and the sandspit known as The Gut moor the "islands" with ever-increasing solidity to each other and to the mainland.

The North Truro Plains

Abutting against the Wellfleet-Truro High Plains on the north are the lower North Truro Plains (*Figure 20*). As we have seen, an irregular 100-foot escarpment separates them; it runs slantingly southwestward from about Cape Cod Light on the east nearly to the mouth of Pamet River on the west. Northwest of the boundary the highest elevation of the North Truro Plains is less than eighty feet above sea level.

The casual Plains of North Truro, although covered with layers of drift, lack the appearance of a true moraine. A few broad, transverse low sags, remnants of meltwater streams, cross parts of it; one of these is followed by Highland Road, North Truro, to the Cape Cod Light. Although there are small kettle-like depressions across the surface, the North Truro Plains hold no ponds. Boulders are nearly nonexistent. Apparently the glacial streams which built up the sands and gravels here did not have sufficient strength to roll boulders. Perhaps by the time they reached Cape Cod, those streams had traveled so far from the east that they had lost their power. For we know that ice lay farther off the present coast in this area than in the area to its south; erosion from the seaward side has shaved more from the Cape here. What is left represents deposits originally more distant from the glacier than those of any other part of the interlobate moraine.

It was these downs of North Truro, then richly forested, that received the first exploratory footprints of the Pilgrims in America. A small group set out along Salt Meadow, southeast of High Head. Soon they came upon "springs of fresh water, of which we were heartily glad, and sat us down and drunk our first New England water with as much delight as ever we drunk drink in all our lives." A saucer-shaped layer or pocket of impervious glacial clay here must have forced to the surface as springs the much-appreciated, cold groundwater which saturates overlying glacial sands. Unfortunately these springs no longer issue; pipes have been laid in the region and take the waters.

Figure 32. *The northern edge of the glacial Cape. This is High Head, North Truro, the blunt end of the Cape's glacial deposits. Here the Cape drops down to the salt marshes and sand dunes of Provincetown.* B. CHAMBERLAIN

Returning refreshed to their ship, the Pilgrim group then sailed south along the low bluffs of the North Truro Plains. When they reached the channel of Pamet River they turned in and explored it for a distance. This led to the discovery and appropriation of the Pamet Indians' caches of corn and wheat on a hill of glacial drift which the Pilgrims named Cornhill, northeast of Little Pamet River.

If, like the Pilgrims, we follow the North Truro Plains to their end, we find ourselves as far north as High Head near Pilgrim Heights. Here, overlooking Provincetown, Cape Cod's glacial substance ends, giving way to recent sea-built sand deposits, marshes and lovely Pilgrim Lake. Before the sea built the land of Provincetown, High Head was a long headland, jutting into the water and bearing the brunt of the storm waves which carved it into a sea cliff (*Figure 32*). It was the termination of Cape Cod. As the sea tore materials from these ice deposits, it built with them much of the sandy foundation of Provincetown, until in time it cut itself off from its source of supply. The building of Provincetown by water continues, but now all of the raw materials must come from down along the coast.

Leaving the road and the marshlands at its foot, we can climb up onto High Head. Before long we are aware of a change in the ground beneath our feet. As we go uphill, the earth becomes dirty-looking, a sticky, muddy brown. Left behind down at the foot of the bluffs are the clean sands of Provincetown, glaring white and yellow in the bright sun—"sand only, so free from all earthy matter that it will not even discolor water," as Whiting said in 1867. Yet some of those very sands were once parts of the brown, earthy High Head bulwark, before the waves raked them down and spread them out to be cleansed. For of all the processes of geological creation, none washes and renews the earth's supplies as thoroughly as the sea.

CHAPTER VII

Reading the Rocks

. . . This Island, as also all the rest of these Islands, are full of all sorts of stones, . . . many of them glistening, and shining like minerall stones, and very rockie. . . .

JOHN BRERETON
Passenger on Gosnold's ship, 1602

Embedded in the earth and clay of Cape and Island moraines lie thousands of boulders and cobbles, adding the spice of ruggedness and irregularity to the terrain. Walking across the land, we may scuff our shoes on them; digging down, our spade will surely strike one; standing on an ocean beach fronting a moraine, we hear their grating roar among the breakers.

The giants in the Cape-Island earth are the huge glacial erratics. They lie concentrated in frontal moraines, those moraines built up along the fronts of ice lobes rather than along the edges where lobes joined. There are large rocks too in the edge, or interlobate, zones of course, but they are scarcer, and rocks are scarcer still in the outwash plains. Wherever the sea makes inroads into the land, such boulders spill onto the beaches, slowly—very slowly—to crumble before the pounding surf.

Here and there among the larger rocks of the beaches we find boulders split into two or more parts which sometimes have moved away from each other a fraction of a foot. Such split rocks usually lie in the zone between high and low tides. A very large one can be seen on the Brewster beach of Cape Cod. Apparently the drag of water, century after century, has set up stresses against the lower surface of such a rock. It has yielded by cracking and, once cracked, has been wedged apart by freezing water. Inland in the fields and pine woods, tree roots and frost can have the same effect, and we find huge rocks parted. One lies conspicuously in Nickerson State Park, East Brewster.

Figure 33. *Glacial boulders on the beach. Large boulders of endless variety are released from the morainal bluffs by cliff erosion.* B. CHAMBERLAIN

From Faraway Places

The array of Cape-Island boulders is endless. Lavish indeed was the icy hand which scattered them nearly everywhere we may look. Although rocks came to this region with all of the ice advances, most of those on the surface arrived in the final ice, the scenery-maker.

It is intriguing to speculate as to their sources. We stand on a beach boulder; we visualize that boulder perhaps torn loose from the sea-pounded coast of Maine—or chiseled from a high crag of the White Mountains—or bulldozed off some peaceful farm valley. And we reconstruct its slow journey, with thousands of other rocks likewise caught up in the swelling enormity of ice, finally to be dropped like so much chaff on this small speck of land. But to pinpoint the actual birthplace of that rock from all possible sources—that takes careful detective work.

In the following discussion of the minerals and rocks which appear on the Cape and Islands, *Appendix B* (Identifying Cape-Islands Minerals) and *Appendix C* (Identifying Cape-Island Rocks) might prove useful as descriptions and aids in field identification.

Tracing Vineyard Rocks

On Marthas Vineyard, rocks are everywhere in the up-island landscape. Heavy boulder concentrations pave the ground. Piles of rocks are dumped in pits. Countless others are made useful as dikes and fences. Many are colorful granitic and volcanic rocks, onetime intrinsic parts of New England's bedrock. If we compare these rocks with rock formations in southern New England, we find that they match most closely those of southeastern Massachusetts and northwestern Rhode Island.

Our Vineyard rock hunt may reward us with semiprecious gems. All along the island's northwest shore are agates. These cherty, concentrically colored pebbles are sometimes six inches across. We must hammer them in half to see the alternating color rings which make agates so attractive when polished. Carried in the Buzzards Bay Lobe, these agates probably came from the region around the Attleboros, Massachusetts. Some may be from Rhode Island; at least they are associated with white quartz of a type found in northern Rhode Island.

Tracing Nantucket Rocks

On Nantucket Island, large rocks are scarce today, as we have seen. But the remaining pebbles provide some clues of their own to aid in locating their source regions and in outlining the path of approach of the glacier.

Atop Sankaty's cliffs, among the pebbles in the till, we find all of these and more: volcanic rocks from near Boston; sedimentary fossil-bearing sandstones from eastern Massachusetts and Rhode Island; pebbly conglomerate resembling that of southeastern Massachusetts; Cretaceous sediments from the coastal plain, including bits of lignite; metamorphic rocks (gneisses and schists) like some of the rocks along the mainland's coast from Newport eastward. Thus the ice that draped Nantucket with its final layer of drift seems to have moved easterly to southeasterly.

Tracing Cape Cod Rocks

Cape Cod's rocks have been the most thoroughly studied of all in the region. To begin with, there are many more of them. Then, too, all three ice lobes helped to make up Cape Cod; the Cape is a complete small-scale model of glaciation in southeastern New England. Cape rocks are the keys to the reconstruction of ice movement in this area.

Even the most casual observer will notice that rocks are most common on the Upper Cape. Here, within the Sandwich and Buzzards Bay moraines, there are numerous fifteen to twenty foot boulders and abundant ten-footers. Never has Cape Cod lacked rocks to annoy the farmer and to aid the builder. On occasion, in fact, the people of this invertebrate peninsula happily have made ironic profit by exporting granite to its birthplace on the mainland (*Figure 33*).

Today, dirt roads thread through the woods to excavation sites where rocks are quarried. All over the Cape these rocks go, for use in breakwaters and jetties. State regulations require that the outside stones on these structures be from two to six tons. Such large rocks are not hard to find. Frequently rocks must be blasted to smaller pieces to be used, and one or two may occupy an entire truck. Abandoned excavations, because of such blasting, are wonderful places to get a good look at the fresh and colorful interiors of these rocks, for on the surfaces of ordinary unbroken rocks only the dull tarnish of weathering usually is visible. (Quarries still in operation, of course, are likely to be blasting and should not be trespassed upon.)

Granitic rocks may be among the loveliest of all, and such are many of the largest boulders. When these are blasted or broken open they reveal a sparkling crystallinity in hues of pink, white, green, black and gray. Many have the clarity of good building stone.

Color and design run rampant in rocks other than granite. There are dark, salt-and-pepper textured diorites; motley conglomerates and dark slate; large quartz pebbles, white and milky or gray and translucent; multicolored gneisses; white and gray marble; green pearly schists; agates; purple compact, solid-colored rocks streaked by crisscrossing veins of white quartz or by colored dikes of other rock materials. Sometimes there are cracks in the rocks, miniature caverns lined with glittering quartz crystals.

The chain of continuity between the stranger rocks of the Cape and the bedrock ledges whence they were torn has missing links enough to infuse the tracing task with challenge. As an indication of the possible trickery of the clues, rock samples taken from the same moraine have been "traced" through the years by different investigators to sources ranging from Nova Scotia all the way to southern Massachusetts.

It is very nearly impossible that all of these tracings are correct. No one ice lobe covered such a wide arc of mainland coast at the same time, and certainly none converged from such far-apart regions to one Cape Cod moraine. Any *large* sampling of rocks from the same Cape moraine came from a much more limited area. It is important that it be a large sampling, for in dealing with large quantities of rocks we may use statistics to clarify the evidence. If we took only two rocks from one moraine and traced them to their birthplaces, we might indeed find that they came from widely-spaced source regions. One could be direct from some bedrock ledge; the other, from an earlier ice deposit on that bedrock—hence, indirectly, from a different birthplace. Thus we must consider only the bulk of materials in occurrences large enough to be of statistical value.

Despite the wealth of conflicting recorded conclusions, let us compare carefully the types of rocks in Cape till with the rock types in various possible source areas on the mainland. Thus we can pinpoint the probable places of origin of Cape ice deposits. This done, we presumably can follow back the trail of the ice lobes across both land and sea to that source area.

Principles Involved in Tracing

When we start our search, we must consider some clues which are of basic importance. They represent empirical geological laws.

To begin with, it is obvious that the *larger* the rocks picked up by the ice, the more likely they are to survive their travels as at least recognizable fragments.

However, if two rocks equally tough and resistant supply the ice with boulders of the same size, the abundance of either in a moraine will depend on how far it was carried. There is an inverse relationship involved; that is, the farther the ice travels, the less will survive of any given rock formation. Far-traveled rocks are much more likely to have been ground to fine bits in the ice mill by the time its debris is dropped. For as an ice sheet moves, rocks move not only with it but within it, ground and shoved and grated together by the stresses and forces set up in the moving ice.

It has been found that it takes only ten miles of ice transport to destroy nearly all rocks except the toughest, those with uniform, homogeneous textures. Such are granite and compact, fine-grained volcanic rocks such as basalt.

Consider what this implies about eastern Cape Cod's rocks, those at Orleans, for example. Let us suppose ice had carried them from the mainland, southeastward, passing to sea from, say, Cape Ann, Massachusetts. Cape Ann is some fifty to one hundred miles from the various parts of Cape Cod. Because of this long travel distance, tough rocks such as granite and basalt should predominate strongly in the rocks of Orleans' Till—little else would be likely to have survived.

But if granite and basalt *do* predominate in the till, that is no proof that they came all the way from the mainland. They may have, of course. But so far removed is the mainland that it is worthwhile to consider the possibility that many rocks came from some closer source now covered by water.

So much for the tougher rocks of the till. Weaker rocks are soft and crumbly or have planes of weakness. Such are shales, limestones, sandstones, slates, flagstones and schists. These would have been ground to bits in the ice long before they had traveled far. Therefore, if we find many of such rocks in the till, we should

Rock type	LOCALITY, WEST TO EAST: *Buzzards Bay Moraine*	*Junction of Buzzards Bay & Sandwich Moraines*	*Sandwich Moraine (West)*	*Sandwich Moraine (East)*	*Interlobate Moraine*
granite; syenite; gneissoid granite; pegmatite	66	60	45	48	17
quartz-feldspar gneiss	1	0	3	1	0
diorite; gabbro; dark gneiss	3	2	3.5	13	7
basalt	3	5	9.5	7	7
light-colored volcanics	4	11	13.5	11	44
schist	3	4.5	5	2	0
slate; shale	trace	2	4	1	2.5
quartzite	11	9	7.5	5.5	5
milky quartz in veins	8	6.5	7	7	7
sandstone	0	trace	1	2	7
conglomerate	trace	0	0	1	0
miscellaneous	1	1	1	1	2.5

Figure 34. *How rocks vary from moraine to moraine. The figures show percentages of various rock types in Cape Cod moraines. It is evident that granite, a mainland rock, decreases from west to east, while light-colored volcanics, abundant on the sea floor, increase.*

look for a definitely nearer source than the mainland. We should conclude that they were torn almost entirely from nearby rocks now covered by water.

Locating Rock Birthplaces

Thus, indirectly, we have defined two broad possible source regions for the Cape's glacial till, depending on the nature of the till rocks: New England's mainland, and the offshore region between Cape Cod Bay and the mainland. These complement rather than eliminate one another. Let us examine each, for more clues.

The New England Mainland

Most of the New England mainland is as solid as a land may be—a complex of ancient granites and other igneous crystalline rocks, in many places long since squeezed and crumpled by earth pressures into tortured, tightly packed arrays of crystals. Interlarding all this are streaks of lava flows, both light and dark. Like a covering skin in places are layers of sediments, mostly consolidated. And of course, overlying all, are the clays and rocks of the glaciers.

Now, we could walk across the three Cape moraines, picking up and examining large numbers of rock chips and pebbles from each. We could make comparisons with mainland rocks. We then should be able to distinguish a moraine, if such there be, built by an ice lobe which traveled almost directly from the nearest granitic mainland with little or no sea floor to cross.

Geologists have already done such walking and sampling and examining and studying. They have found more granite-like rocks in the Buzzards Bay Moraine than in any other (*Figure 34*). Sixty-six percent of the pebbles from Falmouth to Bourne are granitic. More than ninety percent of the large boulders are granitic. Only seven percent are volcanic. There is scarcely any sedimentary material at all. Volcanics are tough rocks and are as likely to survive ice transport as granite; their scarcity in the moraine indicates that they must have been much less common than granite to begin with.

All that granite in the Buzzards Bay Moraine makes it rockier in general than either of the other two Cape moraines. Taking a spade to parts of it can be a most unpleasant experience. Its clay and sand matrix is exceedingly full of cobbles and pebbles. This moraine has more large boulders and great erratics than the Sandwhich Moraine to its east.

To see why this is so, consider one of the characteristics of granite bedrock in this climate. We must go back to the bedrock ledges from which the granite boulders came. Under attack by weathering, granite develops splits. These form along lines of weakness which are parallel to the surface and are spaced farther and farther apart going deeper from the exposed ledge surface. When a series of glaciers passes the ledge, the granite yields along these splits. Glacier number one removes the surface rock. Later glaciers which follow erode larger and larger fragments, because the deeper breakage zones within the rock lie farther apart.

Since the Buzzards Bay Lobe had been preceded by other ice advances across New England, the granite it picked up came loose in large pieces. Earlier ice had removed crumbled, weathered fragments and the smaller pieces which were nearer the bedrock surface. As we shall see, the Buzzards Bay Lobe carried more granite to Cape Cod than either of the other lobes. Thus, it reached the Cape with more large boulders than either of the others. Although the Cape Cod Bay Lobe also was a powerful erosive machine, it left less impressive results than the Buzzards Bay Lobe, which carried more massive freight.

Glaciers of the Wisconsin Stage first stripped some ten to fifteen feet of soil away from New England, and then took most of the weathered and decomposed surface rocks which had lain beneath the soil. Thus, Cape Cod's recessional moraines are in two layers —a lower till layer of fine-grained materials with few erratics, and an upper, bouldery layer. The lower layer represents the dumping of the soils and surface materials which New England lost first. The higher layer is the rocky material torn loose by ice which followed, eroding bedrock already laid bare by the stripping action of earlier ice. Thus, much of New England's good loam lies deep beneath harsher ground on the Cape and Islands.

So thoroughly did the ice do its job of removing weathered mainland granite that in New Hampshire, for example, where several thousand bedrock exposures have been examined, only forty-six have been found to contain any weathered rocks at all. It takes more than the ten thousand years since the end of the Pleistocene to allow deep decay to get a foothold in solid granite.

The Sea Floor

Let us turn now to the sea floor beneath offshore waters, the second possible source area of till materials. This is a region which naturally is less available for study than the mainland. Speculation about its exact nature takes us over the uncertain line between knowledge and theory.

Geophysicists tell us that ridges of granitic rocks continue from the mainland beneath part of the Gulf of Maine. In addition there are certainly volcanic rocks, probably in greater abundance than granite. A belt of volcanics, widening seaward, passes into the sea at the shore of northeastern Massachusetts. This indicates that a ledge of volcanic rocks may well continue offshore as submerged lava beds. How far out it may reach is unknown.

And of course there are sediments. Sedimentary red beds (red sandstones and shales) of Nova Scotia lie partly submerged beneath the Bay of Fundy. These extend possibly as far south as the submarine platform known as Jeffrey's Ledge, in the Gulf of Maine about one hundred miles north of Cape Cod. And finally, any ice which crossed this region would have reaped a rich harvest from the remnants of once-vast Cretaceous, Tertiary and early Pleistocene sediments, which blanket the continental shelf and which already similarly had been eroded during earlier ice advances.

The continental shelf, even where it had remained submerged at the lowest sea level of glacial times, was subject to erosion as far east as one hundred and fifty miles east of Nantucket; thus far seaward were its waters less than six hundred feet deep, less than even a minimum estimated thickness of ice.

Having formed an admittedly rough picture of the sea floor, let us look for a moraine which seems to correspond. In such a moraine we should expect relatively little granite and many volcanic pebbles in the till. Unfortunately, we cannot expect to find a high proportion of sediments because even on a short journey these would tend to disintegrate to unrecognizable bits of sand and clay.

We examine the compositions of the three Cape moraines. And there is one indeed that meets the conditions—the interlobate moraine of the Cape's north-south forearm (*Figure 34*). There is little granite here, only seventeen percent of the stones. Nearly half—forty-four percent—are volcanic. Seven percent are recognizable sediments, and this is a high proportion to have remained undestroyed, unless they came from nearby. Even fragments of coastal-plain fossils have been found in the till of this moraine.

Thus we have accounted for the sources of two moraines, one the land and one the sea. The third moraine, the Sandwich Moraine along the Upper Cape, links these two. It is nearer the mainland than the interlobate moraine, farther from it than the Buzzards Bay Moraine. We should expect its rocks to be a compromise, having arrived in an ice lobe which passed partly over the mainland, partly over the sea floor.

And so the pebbles in the till indicate (*Figure 34*). There is less granite than in the Buzzards Bay Moraine (forty-five percent compared with the latter's sixty-six percent), but more than the interlobate moraine's seventeen percent. Volcanics in the Sandwich Moraine, on the other hand, rank higher than in the Buzzards Bay Moraine (twelve percent compared to four percent), but lower than their forty-four percent in the interlobate moraine. Recognizable sediments show up at two percent in the Sandwich Moraine, another in-between figure for the in-between moraine.

The telltale compositions and textures of the rocks show that the Cape Cod Bay Lobe which built the Sandwich Moraine pushed down along a front that reached from northern Rhode Island to New Hampshire. The granitic rocks in the western part of the moraine made the journey across the present Bay from bedrock of northern Rhode Island and eastern Massachusetts.[13] When the bedrock of these areas is examined, in fact, glacial scratches are found on it pointing toward western Cape Cod. The rocks in the eastern part of the same moraine came from southeastern New Hampshire and northeastern Massachusetts. Striations on the bed-

[13] *The Hingham Boulder Train:* In 1936, Dr. Oliver Howe traced what appeared to be a boulder train from Hingham, Massachusetts, to all over the Cape and Islands. A boulder train is a series of ice-dropped boulders, quarried by a glacier at one locality and dropped in a spreading fan as the ice moved forward and outward in its characteristic mode of flow. A rock very similar to the "Hingham felsite," as Howe called the purplish, compact volcanic rock, outcrops at Hingham. However, other Hingham rocks are conspicuously absent on the Cape, even though they are distinctive, widely exposed on the mainland, and easily recognized in till. Furthermore, the Cape glaciers apparently crossed three different parts of the mainland; they did not cross Hingham and then fan out. Thus, it seems more likely that much of the Cape's "Hingham felsite" comes from a nearer source—the volcanics beneath Massachusetts Bay and the Gulf of Maine.

rock in this region point like a fleet of arrows toward eastern Cape Cod (*Figure 5*).

Its southeasterly route took the Cape Cod Bay Lobe across miles of what is now water. Even if sea level were then as low as the estimated 450-foot glacial lowering, the ice still would have had water to cross. But there would have been more land surface too. In fact, if sea level today dropped that low, the central New England coast would move seaward some score of miles. The ice found these additional miles of land available for easy erosion.

Doubtless each of the moraines can boast stones, scattered among its pebbles or isolated as boulders, which have come all the way from western Massachusetts, Vermont, northern New York and even Quebec. Apparently none have been traced definitely such a distance, although one geologist of the late nineteenth century stated that some of the till pebbles on eastern Cape Cod were "unmistakably" from the Canadian Laurentian region. In other parts of the world, tough crystalline rocks have been found which are known to have traveled as much as 800 miles encased in ice. There is no reason why Cape Cod stones which do not fit into nearer bedrock sources could not have come that far. Some stone may yet be found which mutely will lead us on a journey spanning the miles and the centuries—back to some barren ledge of Labrador, to the birthplace of the glacier itself.

Part Two
ONLY YESTERDAY, TODAY AND TOMORROW

The hills are shadows, and they flow
From form to form, and nothing stands;
They melt like mist, the solid lands,
Like clouds they shape themselves and go.

TENNYSON
In Memoriam A.H.H.

Part Two

ONLY YESTERDAY

CHAPTER VIII

Soils, Sand, and Shifting Dunes

. . . Thence they sailed southward along the land for a long time and came to a cape; the land lay upon the starboard; there were long strands and sandy banks there. . . . they . . . called the strands Furdustrandir *[Wonder Strands], because they were so long to sail by. . . .*

THE VIKING SAGA OF ERIC THE RED
Recorded Thirteenth Century

The Early Postglacial Scene

It might be interesting to pause here and turn our faces away from the harsh years of glacial ice. We will look for four chapters at some of the gentler physical characteristics born of that stupendous sculpturing; we will see how life has taken hold of these exposed but lovely shores; and how men with simple tools and rugged constitutions have made use of the raw gifts of ice and sea.

The warm sun which pushed the glaciers back from offshore New England shone down upon a strange assortment of desolate earth mounds and stream-channeled plains. Shaggy mammals wandered onto the newly released lands, witnessing a scene which we should not recognize today. Long ridges zigzagged across the landscape. These ridges included the Islands' terminal moraines and Cape Cod's recessional moraines. But there were others as well, now lost, perhaps forever, beneath the sea.

If we took soundings across Cape Cod Bay today, we could find two of these ridges, slanting southeastward. They are recessional moraines. They outline the positions of the front of the Cape Cod Bay Lobe as it melted back. The inner ridge, dropped by ice not long after it left the Cape, runs along a line between Barnstable and Wellfleet Harbors. The outer one trends in the direction between the Canal and Provincetown.

As the mammals foraged for tundra plants which in a short time had threaded their roots through the stony till, they may have been able to walk completely around Massachusetts Bay and Cape Cod Bay. They may well have roamed the plain that is now the bottom of Buzzards Bay. At first, only small bodies of water lay in the deeper of these basins, and all probably were cut off entirely from the sea. Low plains bridged the present gaps between Cape Cod and the Islands and the mainland. For we know that during the height of the Wisconsin-Stage glaciation, sea level may have been as much as 450 feet lower; yet today, most of the water which separates the Cape from the Islands is less than a hundred feet deep. To the east at that same early time, land stretched as far as Georges Bank.

Plants and Animals Arrive

The postglacial seasons passed and the climate continued to warm. Tundra animals and plants pressed northward. But other, smaller mammals and reptiles which also had helped to populate the glacial plains remained. These were more adaptable to changing climate conditions, fitted for survival throughout the hundred centuries that by now have passed since the ice vanished.

Some of the low-growing northern plants also remained, and they still live on the Cape and Islands—bearberries, bunchberries and checkerberries, golden heather and poverty grass; leatherleaf and water lobelia. Today these share the fields and woodlands with species from warmer climates. The latter are lingering remnants of plants which came a mere thousand years ago, when there was a period during which the earth's climate briefly reached its maximum postglacial warmth. In many parts of the nearby mainland such warm-climate plants no longer exist, for the climate again has shifted and become too cool for them. But they remain on the more temperate Cape and Islands. We can go to Morris Island, Chatham, and find white cedars in a lovely white cedar swamp; we see hollies growing at Sandy Neck, Barnstable; sour gums, or beetlebungs, form a shady grove at Beetlebung Corner, Chilmark, on the Vineyard; and here and there in the fields grows golden aster.

As the ice melted and retreated ten thousand years ago, the sea, as a result, rose continuously. It filled in the lowland of Vineyard and Nantucket Sounds and connected the large Bays to the sea. In time it made islands of Nantucket and Marthas Vineyard and an almost-island of Cape Cod. It covered the two recessional moraines just northwest of Cape Cod. It lapped against the arm of the Lower Cape, possibly smoothing the surface of the North Truro Plains and the tablelands of Nauset. Almost as effectively as an unscalable wall, it imprisoned the small animals which had established themselves on the higher lands and left them to fend for themselves and adjust to their new, esoteric situation. In this way Marthas Vineyard and Nantucket doubtless acquired many of their little creatures which, hiding out in heath and forest, witnessed the coming of the first white men, and probably the first Indians, to these islands: moles, field mice, shrews and spadefoot toads, box tortoises, the Vineyard's black snakes, and Nantucket's DeKay's snakes.

Through such long isolation, numerous small animals developed their own little idiosyncrasies and became subspecies or even species distinct from their ancestral types on the mainland. For instance, on the Islands there are two subspecies of the tiny short-tailed shrew: the Marthas Vineyard short-tailed shrew and the Nantucket short-tailed shrew, On the Cape and the mainland there is a third species of the same animal. All three started as one species and, through isolation, all three are different animals today. Likewise, the mainland variety of white-footed deer mouse changed into the Marthas Vineyard deer mouse and the Monomoy deer mouse, which is no longer isolated and distinct, since Monomoy has become attached to the mainland during some periods. Nantucket's Muskeget Island has a little mouse unique to it, which originated as New England's meadow mouse. Yielding to its sandy island environment, it changed its habits—learned to burrow among the dunes, to feed on the spikes of beach grass; discovered that when cut, such spikes stay green and juicy all winter if covered with sand. Thus, the meadow mouse became an entirely new species, the beach vole.

Much of Cape Cod's fauna walked or crawled directly over from the mainland via a land connection that was maintained until the building of the Canal. Thus, the Cape itself, excepting Monomoy, has produced no subspecies differing from mainland creatures. Wherever its fauna varies from that of the mainland, it is due not to isolation but to a difference in type of living environment or to local extinction. For example, the Cape has no poisonous snakes. Neither rattlesnakes nor copperheads inhabit the sandy-floored woods of the peninsula, yet they may be found among the rocks of nearby mainland mountainous areas. Copperheads simply do not like sand; they greatly prefer to hibernate under rocky ledges. Rattlesnakes probably did live on the Cape at one time, for they were known even on Nantucket in the eighteenth century, where the sighting of one gave Snake Spring its name. However, they were stamped out years ago from both places by the efforts of local residents. This is true too for the wolves which once roamed Cape Cod and howled across its sandy plains at night.

Land Uplift and Sea Rise

During early postglacial times, as animals and plants spread across it, the land was responding to its new release from the weight of the ice. In the north, where the greatest ice thickness had pressed down on the land, came the greatest uplift when the ice withdrew. Southward, it was progressively less. This process contended with the rising of sea level. The sea level shift, of course, was worldwide, while the pressure release of the land occurred only where the ice actually had lain.

Since we can measure land uplift only against the present sea level, wherever uplift has exactly kept pace with sea level rise it is not apparent that either has changed at all. Somewhere between Provincetown and Nantucket is the line along which such an equality exists in land and sea level rise. South of Nantucket, since glacial times, the land has not moved significantly, but sea level has risen, drowning the margins of the continent. This drowning has occurred north of Nantucket too, but here the earth's crust was depressed as well. Thus, although sea level has risen here as much as to the south, the earth's crust has risen also, and is rising still.

Original Outlines of the Cape and Islands

Thus the land strained to reach stability. The sea approached its present level and filled in the Sounds, and Cape Cod and the Islands gradually took on a more recognizable appearance. We even might have been able to identify a map of them drawn ten thousand years ago, in the uncertain way one may recognize a picture of a grown man taken when he was a child.

As we have seen, where the shores of these lands now face open ocean, the land then reached seaward miles farther, for not only has the sea removed a vast amount of land since, but the water at that time had not yet risen upon the land to its present height. It was perhaps six thousand years ago that the relationship of land and sea became as today, except for subsequent effects of erosion.

In some places, lands have been added since glacial times. Marthas Vineyard, for instance, reached eastward only as far as Poucha Pond and Cape Poge Bay; the waters of Nantucket Sound came in that far. Cape Poge itself did not exist, nor did the beach which now runs southward from it to Wasque Point. There was no South Beach along the ocean shore and no Katama Bay, for the bar ending at Norton's Point and separating the Bay from the Atlantic had not yet been built. There were no long freshwater ponds in the outwash plain; all coves and valleys opened directly into the sea. The southern coast was thus a plain shredded into blunt necks, wide estuaries, and jagged points.

Nantucket Island did not yet have its fine harbor, for there was no Coatue Beach to protect it. On the west, Eel and Smith Points were missing and consequently there was no Maddaket Harbor. On its east and south, the island reached an unknown distance farther to sea and, as on the Vineyard, the present ponds were then saltwater fingers of the Atlantic.

Cape Cod did not yet have its "sandy fist at Provincetown," enabling the state of Massachusetts, as Thoreau pictured it, to stand "on her guard, with her back to the Green Mountains, and her feet planted on the floor of the ocean, like an athlete protecting her Bay." High Head, North Truro, was the end of the Cape (*Figure 27*). Here, as we have seen, where glacial plains now drop down to salt marshes and sandbars, the open ocean then beat against the blunt glacial neck, slicing into being the cliffs which now form its abrupt termination. All the way around to the Outer Beach at Wood End some of this debris was swept. Here it piled up as a shoal, or an island when sea level was lower. Sometimes at low tide these displaced glacial gravels still are evident and doubtless have served as a retaining wall for the building of Provincetown's Long Point.

The Lower Cape was wider at that time, probably by as much as several miles. Nauset Beach was shorter, however, and stopped at Orleans on its south. Monomoy was missing entirely. Barnstable, which now has one of the best harbors on the Cape, then had none of significance, for Sandy Neck was not there to enclose and protect it.

Soil, Past and Present

On these virgin landmasses of ten thousand years ago we can picture the advance of the forests, following the pioneering tundra plants. Forests were certain to thrive in this coastal region with its warm, moist ocean breezes that heralded the death of the glaciers. In the forests, the life cycles of growth, death and decay established themselves. The very first plants to follow the ice had found little true soil. The till itself, of course, had contained a good deal, stolen by ice from northern lands. But this was blended through the debris, diluted and thereby impoverished beyond recognition. It was up to the new, postglacial plant communities to create their own soil, contributing their organic decay substances to the weathered and decayed rocks of the till. The early trees passed on to their offspring the inheritance of soil grown richer with each passing season.

"To know the geology of the [Cape-Island] region," said geologist A. Beaumont in 1941, "is to know its soil." For the development of soil requires rock materials to be weathered first, a very slow process. In this, the Cape and Islands began one step ahead, for their debris consists of many rocks not only weathered chemically but even ground down mechanically before they arrived there.

Such drifted debris, with the exception of areas of thick clay, is porous. Water and the atmosphere can act easily not only on surface grains but even below the surface. Thus, even rock grains which were not weathered when they reached the Cape and Islands were well exposed to weathering agents once they arrived. Over such areas developed the thickest soils. Over compact clayey regions the process was slower and the soils are thinner.

But soil is not only rock material; it is a realm where the inorganic world and the organic world meet and blend. In its formation, plants play an important role. A towering tree falls to the forest floor. Immediately the tiny life harbored there—bacteria and molds—begins its attack. Through the chemical reactions thus brought about, the tree slowly decays and its organic matter becomes humus. This, added to the inorganic components of the ground, creates soil, or enriches soil already there. The humus not only provides nourishment for a new plant assortment, but also holds precious moisture.

For ten thousand years the forests on Cape Cod and the Islands created soil from the stony debris. They worked against formidable odds. Deriving their fragmental material from the till, these soils would necessarily be sandy. More sand was added as beaches formed along shores, for the soils were and still are ever exposed to the sea winds and the inevitable increments of sand blown by them onto the land. The effect of this has been to make the soil in most places so porous that it allows rapid penetration of rainwater. Plant roots must act quickly if they are to benefit from the water. Furthermore, the ever-present winds dry the soil, dehydrating the plant roots and oxidizing organic matter so completely that most of it decays too quickly to form large amounts of humus. Of the humus that does form, much is carried away from the surface by downward trickling waters. Then more soil moisture, lacking sufficient humus to hold it in, is lost to further the fierce cycle.

Had the winds been allowed free sweep over the land ever since it first emerged, the combination of adverse conditions nearly might have prevented any good soils from forming on the Cape and Islands. But the very forests which fed the early soil checked the winds and helped to hold the moisture.

Cape-Island soil today ranges from a few inches to several feet in thickness, although it is more usually thin (less than six inches) than thick. It belongs to the soil classification known as brown podzol, an acid soil common to cool, humid forest regions of southern New England. Cape Cod's podzol is said to be the most well developed in New England. The type of podzol on the Cape is of the Merrimac-Hinckly group—a sandy glacial soil derived mainly from crystalline rocks.

A podzol soil, if left unplowed, usually has a brown surface layer, carpeted with evergreen needles. Beneath this is a gray, light-colored leached zone from which rainwater has removed from the humus the iron, aluminum, calcium, magnesium, potassium and clay. On some parts of the Cape and Islands, only this double zone of soil is present. In many places, however, there is a third, lower zone, darker and richer, for here the elements from the upper are caught and concentrated. Where such soil has remained to complete its development undisturbed, it is rich in humus.

Where this soil overlies such unlikely material as the hard rock-flour clays, it is thin and grades downward directly into clay. But the wide range of elements found in the clay provides the soil with vital plant nutrients so that thin as it is, such soil tends to be fertile. In addition, it holds moisture well, far better than soils of most sandy regions. For example, in 1847 Hitchcock described how a traveler going across the sand plains to the Cape tip would notice a scarcity of vegetation, ever more pronounced—until suddenly and incongruously he would see, in the midst of that seeming desert, a couple of excellent farms in Truro belonging to a Mr. Small, and surrounded by "scarcely anything more of cultivation to the end of the Cape." Mr. Small had either the foresight or fortune to acquire land on top of the blue clay of Truro's Clay Pounds.

Where soil overlies ordinary till it is usually many times thicker and sandier, with rock pebbles scattered through it. In such places there is a complete podzol, with a yellow or brown lower layer. This is more compact and contains the elements removed by leaching from the upper. Cape Codders refer to it as the Vineyarders refer to their hard Montauk Till—hardpan. Rainwater, trickling down from the upper layers, causes tiny clay grains to seep into this zone. Iron oxide descends too, giving the grains their yellow color and mortaring them together. Generally an inch or two thick, this lower soil zone grades downward into the till or gravel beneath.

Even the best of Cape and Island soil is a far cry from the rich loam of interior farmlands. Probably it has always been so. Before the white settlements, the Indians made a practice of burning the forests periodically to clear land for cornfields. When a podzol soil is thus burned over, the friable dark top layer is destroyed, leaving the hard-to-work clay zone. It was thus that the settlers found much of the soil when they arrived. Even into the nineteenth century, farmers fertilized their fields with horseshoe crabs, seaweed and herring, often planting, Indian-style, one fish with each plant set out.

The Plains of Nauset, however, in the course of postglacial centuries, did develop a soil which was rich and fertile, by Cape standards, and which bore good corn and wheat. Early seventeenth-century explorers called the Eastham area a region of "richest soile," with "blackish and deep mould." As we think of production today, of course, the crops were far from spectacular. A Cape Codder did well to produce sixty bushels per acre. Nevertheless, this was a sustenance yield. The wide fields early became known as "the granary of the Cape"; and the sea winds rippled fields of corn which reached from shore to shore.

But farming methods were as wasteful and unscientific on the Cape and Islands as elsewhere in those early days, when there was always more land for the taking. The blockades of the Revolutionary War and the War of 1812 prevented many residents from putting to sea for their wonted fishing, and they turned to the land and overplowed. The fertility of the good earth ebbed away, and erosion finished the job. By 1850, the major part of the granary had been reduced to barren sands.

Before the wastage, the Cape and Islands luxuriated in a green covering never since equaled. Judging from the accounts of early explorers, there was here a forest primeval as noble as any in Evangeline's Acadia. Perhaps we can discount some of the enthusiasm as natural to men long at sea. They were land-starved sailors too, after all, whose wistful imaginings are said to have transmuted sea cows to mermaids. But there must have been some basis in fact for their glowing descriptions of Cape-Island trees.

In 1602, John Brereton, a passenger on Gosnold's ship, described a forest on one of the Elizabeth Islands southwest of Woods Hole, Cape Cod: "high-timbered oaks, their leaves thrice so broad as ours [in England]; cedars, straight and tall; beech, elm, holly, walnut trees in abundance, . . . hazelnut trees, cherry trees, . . . sassafras trees in great plenty all the island over, . . . also divers other fruit trees." And eighteen years later the Pilgrims sailed into Provincetown Harbor and rejoiced to see "so goodly a land, and wooded to the brink of the seas . . . with oaks, pines, juniper, sassafras, and other sweetwood." In fact, the name "Wood End" was given to the neck of land at Provincetown. If it has any significance, it must refer to a forest which then covered this area, ending in the sandspit which juts out from Wood End.

On Nantucket and Marthas Vineyard too, tradition tells of once-rich forests of oak and pine. It was not long before these were stripped to obtain building materials. Even by 1670, logs were being imported to Nantucket and residents were fined if they used

Coatue trees for anything other than housebuilding. For the most part, all that remain of early Island forests are peat beds and occasional tree stumps which, through the years, have come to light along the seacoasts. On Nantucket, for example, stout trunks of fallen oaks lie embedded in the peat.

On the Cape as well, the forest destruction was swift and very nearly complete after the white man came. The trees fell, first to clear lands, then to build houses and ships—the United States naval store industry sprang during the seventeenth century from Cape Cod pitch pines—then to feed the fires of local industry and the frequent uncontrolled fires of human carelessness. By the mid-nineteenth century most of the forests were gone. A few further decades of plowing, harrowing and grazing exhausted the soils—imagine the effects of some 13,000 sheep on little Nantucket Island at one time! Unchecked winds meanwhile diluted the rich soils with sand, dried them out, blew them away.

Of the Cape Hitchcock says in 1847: "as we approach the extremity of the Cape, . . . in not a few places it would need only a party of Bedouins to cross the traveler's path to make him feel that he was in the depths of an Arabian or Lybian desert." Nine years later, Thoreau speaks of "a thin layer of soil gradually diminishing from Barnstable to Truro . . . with many holes and rents in this weather-beaten garment, not likely to be stitched in time. . . ." And, recounting an old 1802 description of Eastham, that earlier region of "richest soile," he said that one could see a "beach on the west side of Eastham . . . half a mile wide and stretching across the township, containing 1700 acres on which there is not now a particle of vegetable mold, though it formerly produced wheat."

And of Nantucket Island about the same time, Herman Melville gives a description in *Moby Dick:* "Look at it—a mere hillock, an elbow of sand; all beach without a background. There is more sand there than you would use in twenty years as a substitute for blotting paper. Some gamesome wights will tell you that they have to plant weeds there, they don't grow naturally; that they import Canada thistles; that they have to send beyond seas for a spile to stop a leak in an oil cask; that pieces of wood in Nantucket are carried about like bits of the true cross in Rome; that people there plant toadstools before their houses, to get under the shade in summer-time; that one blade of grass makes an oasis, three blades in a day's walk a prairie; that they wear quicksand shoes, something like Laplander snowshoes. . . ."

And on Marthas Vineyard in 1888, Shaler reports that apart from the region, on the west, of local soil enrichment by underlying Tertiary clays, plowing and forest fires had rendered 33,000 acres east of Tisbury untillable. This is fifty-three percent of the land, a land once covered with forests. "The greater part of this land," averred Shaler in those days before the summer sunners, "is not at present valued at more than two dollars per acre, and much of it could probably be bought for a less price."

Trees and Forests

Then, born of desperation, came the earliest recorded reforestation effort in America. Chatham, Cape Cod, was the pioneer. In the early nineteenth century the town lay endangered by Great Hill, the sands of which were actively drifting over the town. In 1821, beach grass and pine trees were planted on the hill to hold the sand. They were successful.

But several decades would go by before the now-omnipresent pitch pine really arrived on the rest of Cape Cod and thence spread to the Islands. For Cape pitch pine seeds Nantucket once willingly paid twenty dollars a bushel. Thousands were set out in rows across the barren fields in a desperate attempt to hold the blowing sands and to restore the one-hundred-century accumulation of soil. Thoreau watched some of these early plantings and mused that "perhaps the time will come when the greater part of this kind of land in Barnstable County will be thus covered with an artificial pine-forest." He was a good prophet. These trees, rapid-growing, generally wind-resistant and tolerant of poor soil, have taken over the Cape with their wide forests and have added to the wild beauty of seaward views. In addition, thousands of the more sedate white pines have been provided by the state, particularly for planting on the dunes.

On one of the Elizabeth Islands there is a lone fragment of New England's virgin hardwood forests, surviving through the centuries like a page from the past. Here on Naushon Island grow red maples, black and yellow birches, hickories, beeches, white, scarlet and red oaks, white and red cedars, and pitch and white pines—"the only proved climax oak-beech forest surviving in New England," according to the National Park Service. On the Cape, a small island in Mashpee's Wakeby Pond has a similar tree covering, although early logging may have occurred here, making today's trees a secondary growth. Elsewhere in this area, the stock of trees which are descendants of these firstcomers is diluted by many more recent immigrants.

Across the Cape and Islands today, twisted scrub oaks thicken the forests, growing as kindred souls with the rugged, ragged pitch pines. There are stands of black oak and red maple in these forests as well. Other hardwoods shade homes in the towns. Beech, willow, locusts, white oaks, gray birches, sour gums and nut trees are common to the Upper Cape and to western Marthas Vineyard. Here and there are orchards with tiny, gnarled apple trees; Sandwich once had a cider press. Lining the streets of many of the towns and particularly notable at Yarmouth, Cape Cod, are majestic cathedral elms, elegant memorials to the seafarers who planted them. On the wilder stretches of rolling sand plains, such as the Lower Cape and Nantucket, pines along the coast fight a near-losing battle against excessively high winds, salt spray, and depredations of the pine moth. Dwarfed and ragged, they nevertheless manage to hold their own, fifty-year-old trees scarcely higher than twenty feet.

Nantucket, with its elusive water supply and exposed location, never could compete with the Cape and Vineyard for trees. But where large trees are rare, small trees will shade nicely, and wild flowers run rampant. However, wooded areas have not been entirely unknown to the island. Such names as "The Woods" and "Grove Lane" for areas now nearly treeless must indicate that they once had trees. Fresh peat and thick trunks occasionally are exposed along the island's shores, as we have seen. High tides cover some of these peat beds today, indicating that the drowned forests grew at a time when sea level was lower and the island larger, presumably just after the ice had left. Trees had a much better chance then. The greater expanse of land modified the winds and climate to some degree. Trees which established themselves got a good grip on the soil with their roots and prepared to stay. As the island shrank with the rising water, those trees which were not right on the coast should have survived until today, adapting themselves to the harsher conditions. But the trees went when the settlers came.

Other trees have come to Nantucket, however. Since 1831, trees have been planted in Nantucket Town and have fared well, thanks to loving care. Century-old elms continue to pattern the cobbled streets with their long shadows. Outside the town there are areas where the soil is not all sand but is enriched with dark vegetable matter. There is such a place near Sankaty, and here grow such hardwoods as oaks. In the wilds, however, and along the coast, the soil is sandy and bare; winds are strong; and mortality is high indeed.

Over most of the Cape and Islands today, not even our great grandchildren will see the soil restored to its original thickness. But still it is gradually forming. In the sheltered recesses of the pine forests, blowing sands are controlled. Pine-fringed fields are harbored from erosive winds.

Western Marthas Vineyard has retained its fertility well. It has been said that in West Tisbury grows every type of wild flower found in Massachusetts. Forestdale on Cape Cod, north of Camp Edwards, is another fertile region. This has a deep water table. Builders and land clearers have shied away in favor of places where shallower wells may be drilled. Thus, Forestdale has remained relatively undisturbed through the years, and beneath some fifteen hundred acres here is good soil nearly twenty inches thick.

Farming Today

Today, considerably less than a third of the Cape and Islands is used for crops. This is far below their potential. Of Cape Cod, for instance, an eighth is morainal, rocky and hilly; another eighth is barren sandspits, dunes and beaches. The remaining three-fourths could be tilled. Of this, of course, much is now covered with pine and oak forests and broken by ponds and swamps; and much is rapidly giving way to the structures of civilization.

The small but varied crop of vegetation which the soil supports enriches and secures it. The very sandiness of the loam makes it excellent for asparagus, for which Eastham was once famed and which still grows, sometimes wild, across the plains. Cape Cod's 700-acre strawberry industry, centered in Falmouth (where as much as 15,000 quarts per acre have been produced), also owes its success to the sandiness of the soil; as do the potatoes, truck gardens and poultry farms of the Cape outwash plain. On the Vineyard, cattle, horses and sheep graze in the lush pasturelands of the west, for the soil here is nourished and kept moist by preglacial sediments beneath it. Most of Nantucket, on the other hand, is happily content to raise roses and to farm the crops of the sea.

But we have not yet mentioned the most important crop of all, on all three lands. We must turn from the fields and look to reclaimed swamplands to see it—the cranberry. Today Cape-Island cranberry bogs, together with those of neighboring Plymouth County, produce half of the entire cranberry crop of the nation. At the turn of the century, when much of the land lay wasted and barren, these bogs were the aristocrats of real estate. They found buyers at one hundred dollars an acre, while other unwooded lands sold, "in the rare transfers . . . effected," for about fify cents an acre.[14]

Cape-Island Sand

As we have seen, the Cape and Islands have always been battlegrounds where sand is the attacker and trees the defenders of the soil. The long struggle began when the first waves broke against crumbling banks of morainal drift and created the first sandy beach.

Wherever there are agencies powerful and persistent enough to disintegrate rocks, sand will appear. In some deserts, for instance, arid climates hinder plant growth; winds are unchecked; rocks crumble into wind-borne fragments which smash against each other and the ground until sand dominates the land. And not only in deserts, but everywhere, the weathering and decay of bedrock and large boulders causes component grains to fall out as sand. Where such rocks lie as till boulders or as bedrock along the earth's shores, the sand grains become prey to the breakers of the open sea, which dislodge them from their resting places and throw them to the beach below.

This takes us of course to the Cape and Islands. Here the shores consist of vulnerable soft banks of glacial drift. The sea's inroads are rapid, and sand forms in abundance. Much of it comes ready-made from the till itself. Even where the bodies of the Cape and Islands are not morainal, as, for instance, the Vineyard's Cre-

[14] For more about cranberry bogs, see pp. 190–191.

taceous and Tertiary headlands, they are made nevertheless of sand and clay. These are easy grist for the sea. And the mill of the surf is relentless and constant.

Once they are spread upon the beaches, the loose sands are never stationary. They move with the wind-driven water. They pile up as sandbars, spits and hooks. Beaches, in fact, are simply left-behinds, surfaces from which sand gets endlessly removed and replaced. In the processes of land construction and demolition they are geological loading platforms. Within a matter of months a beach may increase or decrease in height by ten feet.

If we stand on the sandy beach in front of the cliffs at Gay Head, or at Siasconset, or at Cape Cod Light, most of the sand grains at our feet will have come not from the cliffs behind us but, carried by coastal currents, from somewhere down the coast. With the sea winds themselves sand moves, carried inland from the beaches and piled up as dunes. The dunes move too, if unhindered, in a smothering invasion of the land.

Not only are sands in endless motion from place to place, but they are continually subject to change by the forces which created them. Few grains remain permanently on a strand. If they do not become part of a dune or a sandbar or a new island, they receive the attack of other moving sand grains surrounding them. Beach sand grains are likely to be coated with thin films of water, serving like auto bumpers to protect them from the shock of collision with one another. Nevertheless, long and relentless buffeting takes its toll. Eventually they may reach so tiny a size that the waves can sweep them up and support them, and they drift out into still, quiet waters. Here they settle as mud, feeding the tidal flats.

Astronomical indeed is the number of sand grains on the beaches, and each one has a long and complex history. It goes back to the birth of the rock from which the sand came. For there is no sand that was not at one time an intrinsic part of a solid mass of the earth's crust. And to go even further back to ultimates, there is none that was not a liquid part of molten upwellings from subterranean sources.

Since a large part of the sand grains of the Cape and Islands came directly from glacial drift and thus indirectly from the complex rocky surface of New England, these sands claim as ancestors representatives of nearly the whole spectrum of rock types to be found in the earth's crust.

A Close Look at the Sand Grains

If we were given a handful of sand and asked to determine whether it was from a beach, a dune, or from glacial till, we could make a pretty good informed guess. Beach sand grains are relatively large; their edges are smoother; and those found together in any one place are of rather uniform grain size. Dune sand grains are smaller and more nearly round in shape. Their surfaces may be frosted from wind blasting by bits of grit. There are more tiny grains of silt among the sand grains. The sand of glacial till, if left undisturbed where it was dumped, tends to consist of many-sized, irregular grains with rather angular edges.

Let us look at a handful of beach sand under a strong magnifying glass. Although Cape-Island sands have minor variations from shore to shore, there are basic similarities. The grains in our handful, in accordance with the geological definition of sand, are from 1/500 of an inch to an inch across.

Reference to *Appendix B* (Identifying Cape-Island Minerals) should prove helpful in identifying the constituents of our handful of sand. Most of the grains are white or colorless, glassy, and nearly round or egg-shaped. We can scratch glass with one of these. They are the mineral silica, or quartz—exceedingly common in the igneous and metamorphic rocks of the northeast from which Cretaceous seas and Pleistocene glaciers each drew their sediments. Quartz is tough and inert. Nature can affect it most noticeably only by a mechanical means of attack. Winds, waters and glaciers can tear quartz from its original rock body. They can buffet it and grind it, vent all their forces upon it, and succeed only in breaking it into smaller pieces. Thus, it is the mineral most likely, of all common minerals, to survive and make sand.

The other minerals which, with quartz, are commonest in New England rocks have survived only poorly the rigors of the earth processes which have placed sand upon the Cape and Islands. These other minerals are more vulnerable to the chemical attack of the atmosphere. Oxygen, water and carbon dioxide reduce them to clay or rust or to an array of other soft materials. In fact, in many cases it was their destruction in their original rocks that freed the quartz grains from those rocks and set them on their way to sand deposits.

So thoroughly has nature refined and concentrated quartz from New England's mixtures of minerals that on the Cape and Islands, quartz makes up nine-tenths of the beach sand grains. The remaining tenth is no less interesting, however. It embraces the minority of rock minerals that thus far have escaped chemical and mechanical destruction.

If we look more closely at the sand under our magnifying glass, we see it is lightly peppered with dark grains. We place some sand on a sheet of paper and run a magnet under the paper. Some of the dark grains walk with the magnet. These are probably magnetite but may be ilmenite. Magnetite is a mineral of iron oxide. It is ubiquitous although not abundant, occurring in very small amounts in nearly all of the rocks of the earth's crust, and in large amounts in some. Ilmenite is an iron, titanium oxide mineral, far less common than magnetite. These two minerals are particularly well represented in the Vineyard's preglacial deposits, so that they are common too in the beach sands derived from Vineyard cliffs. In some places such sands have been reported to be twenty percent magnetite and ilmenite. Such a concentration makes the sand appear dark. At one time, the beach at the foot of the Nashaquitsa Cliffs and that reaching from Vineyard Haven to West Chop Light were composed of such dark sands. In 1888, Shaler suggested their possible use as iron ore, although their concentration was not sufficient to make the attempt profitable. Today, rock jetties along the beach deflect other sands, water-carried from farther away. These have covered many of the black sand beaches, hiding the iron minerals from view.

Let us turn to the rest of the minerals in our sample of beach sand. We see greenish-black grains, frequently longer than they are wide. They are likely to be the mineral hornblende, a complex compound of calcium, iron, magnesium, aluminum, silicon and water. There is some mica too, in glittering paper-thin flakes, black, bronze or colorless. Some feldspar may be present, light in color (tan, white, gray or pink), dull and blocky-looking in contrast to the glassy grains of quartz. Tiny dark red or pink grains indicate garnet, and round green bits may be olivine. All of these, like quartz, are common rock minerals. They make up most of the tenth of the sand that is not quartz.

There may be grains of more exotic minerals too, of course—pink rose quartz; purple amethyst (quartz); pink, green or black, long, shiny crystals of tourmaline; yellow or colorless topaz, harder even than quartz—and perhaps, if we are lucky, even beryl and sapphire. We may be amazed to discover what we have been crunching underfoot.

The wind sometimes refines these more unusual minerals, concentrating them as layers in the sands of dunes. We look at the sloping face of a sand hill and may see streaks of dark and light alternating across it. The light is probably quartz. Garnet or magnetite, heavier minerals, are likely to make up the dark bands. Winds which can move quartz grains may be too weak to move heavier minerals. The quartz is then blown from the beach, while garnet and magnetite are left behind. The quartz settles to form a light-colored layer on the dune. Then a stronger wind may come, picking up both light and heavy minerals from the beach, spreading a layer of the mixture above the pure quartz layer in the dune. Now if a weak wind strikes the sand hill, it can move only the quartz from this mixed upper layer, whisking it away to leave behind dark minerals.

Processing of Beach Sands

Every time we watch the waves of the sea pounding against a shore, we are watching a sand factory at work. But, for a thrill, let us stand on the cliff by Cape Cod Light, North Truro, during a northeaster, when the tide is full and the lashing of the waves competes with the roar of the winds to fill the air with tumult.

Under such conditions, the beach vanishes altogether. Breakers hit the cliff face with a force up to several tons per square foot. The earth materials of the banks fall; they become part of the churning waters and are sorted into sizes. Most boulders remain at the foot of the cliff. Cobbles and pebbles fall back with a hiss and a clatter to a level between the tides where, endlessly ground and polished, they will stay to tease the feet of summer bathers. Other pebbles remain scattered on the beach, at least until the next storm. We see such stones, where they have been undisturbed since the last tide, lying like sheep, all facing the same way, their lengths parallel to the shore. Sand slips farther seaward to be caught up in the current which flows along the shore and laid down upon some other beach, some other time. Silt and clay drift away to sea to settle in deeper, more quiet waters.

Even along the shores of more protected water, such as Cape Cod and Buzzards Bays and Vineyard and Nantucket Sounds, beach sands form, raked and pulverized from scarcely resisting banks of till. Now, after centuries of such sand formation, Marthas Vineyard and Nantucket lie like green pictures, mounted in the sea with wide, white frames of sand. The sand beaches of Cape Cod could be laid out to reach from New York City to Baltimore and back again. The thirty miles of gleaming beach from Monomoy to Provincetown on Cape Cod—the great Outer Beach of Nauset—is thought by some, on convincing evidence, to be the Wonder-Strands (*Furdustrandir*) of Norse sagas. To the Vikings, arriving from rock-rooted shores, such a long, long beach—sometimes, depending on the mood of the sea, as much as four hundred feet wide; sometimes less than a dozen—surely must have appeared wondrous; for it is wondrous even to those of us today who have seen it many times. When Samuel Champlain traveled to this coast on an early voyage, the bright sands held his attention too. He named the peninsula which is now Cape Cod, *le cap blanc* (white cape), "because there were sands and dunes which appeared thus."

Uses for Cape-Island Sands

In view of the plentiful supply of sand, it would seem natural that Deming Jarves selected Cape Cod for the glass factory which became the home of the Sandwich Glass Company. In Sandwich there was as well proximity to a seemingly endless supply of red oakwood from the forests and of salt creeks for shipping products across the marshes to and from Cape Cod Bay. The factory was built, and in 1875 the first glass was blown, using Plymouth sand, much like that of the Cape. It was a sad failure. Such sand apparently had a vital fault, all that iron in the form of minerals from the glacial till. So, to this land of inexhaustible sand, more sand had to be shipped on barges from New Jersey and western Massachusetts. There was nothing wrong with the wood of the forests, however. So much went into the furnaces for glass production that the Sandwich Glass Company played no small part in their depletion.

The grittiness of Cape sand brought it fame in another industry. Building-stone dealers of nineteenth-century New England preferred it above other sands as an abrasive for polishing and sawing granite and marble. So great was the demand that Provincetown dunes were leveled to fill it, ships carrying the sand to Boston to return laden with badly needed soil. The less refined interior sands made excellent mortar; building contracts in east coast cities often specified "Cape Cod sand."

Cape Cod Dunes

No sooner does sand create beaches, either bordering ice deposits or as current-built spits and bars, than dunes begin to raise themselves (*Figure 41*). It is not surprising that in some areas, dune mountains and dune ranges are the most dramatic elements in the scenery.

The entire curved spit at Provincetown is composed of dunes, old and new. Among them are the highest on the southeastern New England coast. Pilgrim Monument stands on a dune, Town Hill, nearly a hundred feet high. In the wild and beautiful country behind the town, many of the dunes are forty to fifty-five feet high. The old, outermost line of these dunes, Peaked Hills, is crossed by the Race Point road which, running from Provincetown to the Race Point Coast Guard station, takes us through some eight square miles of magnificent scenery. This dune range reaches from Race Point along the Cape ocean side all the way to the glacial deposits at High Head, North Truro.

South of Peaked Hills and separated from them by Race Run (a low valley parallel to the coast) is a second line of dunes. These are more dignified, higher by about forty feet. They run from the two high Biblical dune hills west of Pilgrim Lake—Mount Ararat and Mount Gilboa—to within a mile of Hatches Harbor at Race Point.

The dunes of Provincetown itself, the inner dunes, are motionless today. They will not walk away with the town on their backs. They have been trapped in place by a net of roots, the roots of more than 500 species of plants. Many of the dunes outside of town, however, particularly the western dune ranges and those spectacular hills of shiny pristine sand now lying between Route 6 and the ocean, are very much in motion, despite efforts to hold them down with beach grass. Before the north winds of winter, many thousands of tons of sand sweep annually from windward to lee sides of these dunes. When we travel past them, we see their surfaces pierced by the stark trunks of dead pines and the branches of live ones, low, twisted and gnarled. "Indeed," wrote Yale president Timothy Dwight in 1822, "a Lilliputian of three score years and ten, compared with a veteran of Brobdingnag, would very naturally illustrate the resemblance, or rather the contrast, between one of these dwarfs, and a full-grown tenant of our forests."

These dwarfs are the tattered remnants of pitch pine woods that had stood in the path of the rolling waves of sand (*Figure 35*).

Figure 35. *Sand dunes engulfing trees. If these Provincetown pitch pines can keep their crowns above sand, they may survive.* B. CHAMBERLAIN

Those trees which became completely covered perished. Later, when the dune moved on, first the tops and then the bare trunks of the dead trees came to light. Even though an unbroken white sea of sand today may show no sign of trees, if we dig to twenty feet down we may find soil and the charcoal-like stumps of pitch pines. Layers of buried soil are not uncommon, in fact, and often stick out horizontally partway down the dune slope. The live trees which struggle out above the dunes are those which have managed to keep their crowns in the air and thus to survive partial burial. What we see as stunted dwarfs a yard or two high are likely to be actually the tops of full-sized trees.

The dunes of Provincetown are the largest and most magnificent, but are by no means the only dunes in the Cape-Island region. Smaller than these moving giants are the dunes which rim the ocean beaches from Eastham southward (*Figure 38*), where the morainal cliffs diminish in height and finally vanish. Rows of shifting dunes ripple over Monomoy. Across the Cape, Sandy Neck in Barnstable is covered with large old and new dunes which frame the harbor. Sandy Neck Road takes us through them, and we may see what Timothy Dwight saw—"a long, lofty, wild and fantastical beach, thrown into a thousand grotesque forms by the united force of winds and waves." Elsewhere, up and down the Cape Cod Bay coast there are small dunes huddled behind the beaches. Along Route 6 from Sagamore through Barnstable, we get some fine distant views of these bay dunes, which seem to fence the entire length of shore.

Marthas Vineyard Dunes

The Islands too have their dunes. On the Vineyard, a fine body of them lies west of Menemsha Inlet, in the eastern part of Gay Head. Here the sand hills have journeyed inland a short distance and have buried stands of vegetation. From Gay Head's cliffs to Squibnocket Beach in the southwest part of the island, there are dunes too. These creep inland, driven by the prevailing southwest winds of summer. Already they have cut off the mouths of several streams, damming them and forming the poorly drained ponds and swamps of that region.

Nantucket Dunes

Low and refined are Nantucket's dunes, like Nantucket itself. But among the low dunes along Coatue Beach and Great Head, clusters of gaiety break forth in July, when the prickly-pear cactus blooms. Now rare, these plants grow at the eastern limit of cactus plants in the United States. The island's at sea location moderates its climate very much, allowing such cacti to grow not far from Scotch heather and the northern creeping snowberry of some of the bogs. It has been said, in fact, that Nantucket has a larger assortment of plant life than any other United States locality.

Dune Growth and Life

Among the dune colonies on the Cape and Islands, all stages of dune existence—birth, growth, movement and death—are evident. The sun dries sands that the sea has carried from one beach to another. Onshore winds snatch up the finer surface grains, winnow them into separate sizes, and sling the smaller particles inland. Apparently, ordinary winds rarely can carry sand grains greater than 1/25 of an inch across; and these not more than six feet above the ground. At this level, wind does not get very far with its load, for it hits obstacles such as brush or trees and must drop the sand grains. A pile of sand thus begins to accumulate. As the pile grows it may stop later winds, which add their loads. Most of the increments of sand which thus arrive blow up the gentle front slope of the maturing dune, reach the crest, and drop beyond it onto the steeper back slope.

Thus, along any beach not banked by high cliffs, a row of low foredunes tends to develop. These are familiar sights on open beaches such as those of Sandwich, Cape Cod. If the prevailing coastal winds come from a direction such as to feed the foredunes constantly with more sand, a series of such wave-like ridges develops. If unstopped, these migrate inland by continuous growth on their leeward sides.

Dunes may form on top of high cliffs too, but such are lowlying and transitory. The winds which lash a cliff down at its base pick up sand and dust. This debris is swept upward by peculiar vortical air movements created by the presence of the cliff. "If you sit on the edge of the cliff," observed Thoreau, "you will have ocular demonstration of this by soon getting your eyes full." In time, tracts of sand and dust pile up on the cliff top. Sometimes these can be seen along the bluffs behind Nauset Beach, or at Nashaquitsa. Part of the sand of such dunes comes directly from till, and the dunes are dustier than piles of ordinary beach sand, for beach sand quickly loses its dust to the sea before it becomes wind-borne and piles up as dunes. On a cliff top, sometimes as much as several feet of windblown sand and dust build up, only to be swept down off the cliff and to sea when a wind from the land comes bustling along to tidy up. Saucer-like hollows may then remain atop the cliff, lined with left-behind gravel; these hollows may localize new sand accumulations.

Nothing looks more inhospitable to life than an area of sand dunes, but life is there nevertheless, and even may be abundant. The sunbaked, waterless upper layers of sand in a dune can reach a temperature of more than 150° F., but beneath the surface we may find moisture from within a few inches to a foot down. Beneath a dune, as beneath any other land, there is a water table, a level below which the ground is saturated with water. Rainwater which falls on the surface trickles rapidly down to the lower layers, and these retain their water supply even during periods of drought.

Even the dead, buried trees in a dune are put to use in the thrifty-profligate world of nature. They serve as reservoirs of water which supports colonies of one-celled plants and animals and of insects. Occasionally, in fact, what seems to be the ultimate in the unlikely occurs, and a mushroom sprouts forth into the alien world of blazing sun and white sand, rooted in and fed by the decayed wood beneath. Thus, the provision of water, however scanty, creates a possible plant environment. And plants take advantage of it.

Naturally, when a dune is first blown into existence beyond the beach, it is just barren sand. Then plants begin to appear. The dunes of the Cape and Islands support all stages of plant populations, from the earliest pioneering grasses to a climax stand of trees and the accompanying transformation of dune into ordinary sandy forest bed.

In areas where the winds take a holiday for one reason or another, new sands no longer reach the dune. It is deprived not only of sand, but also of the thin films of moisture which would arrive with sand grains. Such dunes stay loose and dry, and are most readily subject to wind erosion. Although some plants may settle on them, they are not easily stabilized by a growth of vegetation.

Beach grass is the first plant to take hold in this sort of dune, sending its thin roots several feet deep to tap any leftover water, while on the surface it delicately traces circles in the sand with its arching, breeze-blown blades. But unless new sands arrive, this grass is short-lived. In a few years it exhausts the small supply of nourishment from the sand in which it is rooted.

If new sands continue to arrive, however, in small, regular increments, the grasses will survive. They rise to the situation by forming new rootstocks and thus keeping themselves at the surface. Sometimes mossy poverty grass joins the beach grass. Together they provide a slow accumulation of humus.

If the dune's growth is slow, this paves the way for wild cranberries, bayberries, beach plums, and wild roses which cautiously venture out from inland wooded areas. In such regions as quiet inlets, where small fresh supplies of sand are washed in daily by waves, a richer assortment of plants forms immediately on the moist, compact dune surface. Their presence helps to retard inland blowing. Just above the tide line such beach plants as dusty miller, beach peas and sandburs quickly take root. Beach grass follows, along with seaside goldenrod, sand wormwood, bayberries, wild roses, and beach plums.

A rapidly growing dune will reach a height where the free winds are strong enough to tear loose exposed vegetation. Thus, breaking the chains which have begun to bind it, the dune becomes part of a dune range. It may then move inland, smothering trees in its path until it has gone so far that winds no longer can reach it to destroy its plant covering. Then it becomes stabilized.

At this point, the long process of humus enrichment begins. Seeds of pitch pine from neighboring woodlands are carried to the dunes. Here humus from beach plants provides the baby trees with nourishment and established plants give protection. Following the pines come oaks and sometimes beech, with minor amounts of red maple and birch to dapple the sandy hills with autum loveliness. In most regions, however, conifers remain dominant, springing up from a carpet of inkberries, huckleberries, greenbrier and bayberries, poison ivy and woodbine, arrowwood, hazelnut and grasses. This stage comes a century or two after the dune is stabilized. And thus ends the dune. It may keep its hilly outline, but its sand barrenness is now root-tied soil, and wind no longer can bully it away.

Shifting Dunes

Advancing sand dunes once were a serious problem—indeed, a grave danger—on Cape Cod, especially at Provincetown. For early inhabitants destroyed the forests and other vegetation which once grew upon the dune ranges in the town and on its outskirts.

Settlers who came to Provincetown drove their cattle onto the upland pastures that capped the dunes. They did not replace plants which were destroyed, and in time, the pastures were denuded. Farmers cleared woodlots with fire. They cut down forests to build ships and salt factories, to obtain pitch and turpentine, to burn for fuel. With the disappearance of plant communities balanced by centuries of development, and nothing left to hold it, sand began again to blow. It blew toward town and harbor. The government built a beach ridge across the harbor to keep it from filling up, and planted it with beach grass to hold it. Farmers spread their marsh hay to dry on top and killed the beach grass. When far-sighted legislators, early in the eighteenth century, passed an act forbidding tree cutting and another forbidding pasturage of stock on sand hills, there were some who ignored them almost entirely.

The realization of what they had started began to dawn on the inhabitants in the early seventeen hundreds. With their axes, spades and pitchforks they had indeed opened Pandora's box. By 1725, Provincetown was becoming a miniature Sahara, Truro was following close, losing more valuable lands each year, and even Wellfleet could feel the sting of biting sand.

First, Provincetown lost her soil. Then houses became banked and finally buried by sand. Thousands of tons of it had to be carted away annually. Streets were ankle-deep in soft sand. Thoreau's description, written in the mid-ninteenth century, draws a vivid picture:

> The sand is the great enemy here. The tops of some of the hills were inclosed and a board put up forbidding all persons entering the inclosure, lest their feet should disturb the sand and set it a-blowing or a-sliding. The inhabitants

are obliged to get leave from the authorities to cut wood behind the town for fish-flakes, beanpoles, pea-bush, and the like. . . . The sand drifts like snow, and sometimes the lower story of a house is concealed by it, though it is kept off by a wall. . . . There was a schoolhouse, just under the hill on which we sat, filled with sand up to the tops of the desks, and of course the master and scholars had fled. Perhaps they had imprudently left the windows open one day, or neglected to mend a broken pane. . . . In some pictures of Provincetown the persons of the inhabitants are not drawn below the ankles, so much being supposed to be buried in the sand.

People were spared a view of the havoc from their windows, at least, for glass had become opaque from sand frosting.

At the turn of the nineteenth century, the giant dune ranges themselves were moving toward the town and harbor at ninety feet a year. Perhaps it was the danger to their all-important harbor—the very core of their livelihood—that rang an alarm sufficiently loud for the townspeople to heed.

In 1825, beach grass plantations were made on the naked dunes and all grazing stopped. Later, the planting of pitch pines spread to the area. However, by that time the dunes had become hard to stop. Even half a century later, a Trustees of Public Reservations report on the area estimated that more than a million tons of sand from northern dunes still pressed south each year, driven by north winds. Southern dunes, however, had become stabilized.

To save the harbor, a dike of earth and beach grass went up across its western corner from Long Point to the tip of the town. A long, massive rock dike since has taken its place and now protects the harbor from the high seas which once threatened to break across, carrying a cascade of sand into the harbor (*Figure 42*).

Truro was waking up too when Timothy Dwight visited it. It was sandy and barren then, but Dwight noted that the town had impressed upon its inhabitants the importance of planting beach grass three feet apart in rows, to tie down their restless narrow land.

CHAPTER IX

So Necessary to Life

Cranberries and cranberry swamps—or bogs, if you like, . . . are almost as common on the Cape as potato patches and, generally speaking, ever so much more profitable. The flat, green acres, intersected with ditches and with dikes at their ends or sides, are a part of the landscape, and Barnstable County would not be Barnstable County without them. In the winter their flooded, frozen surfaces make good, outdoor skating rinks; in the spring their expanse of glistening vines are like great carpets spread in the hollows; in the summer the billions of tiny young berries are showing, and people speculate as to the size of the crop . . . ; in the autumn is the culmination—profit or disappointment—for the fall is picking time.

JOSEPH C. LINCOLN
Cape Cod Yesterdays

Always, the sea has been the daily bread of the Cape and Islands, but their freshwaters are their jewels. Like gems, hundreds of ponds sparkle from settings of green forests, russet fields, or golden sands. Some lie in solitary beauty, others in casual clusters, still others threaded together onto chains by narrow brooks. They are the surface glimpses of a most precious mineral which lies in rich abundance beneath the Cape and Islands—sweet, fresh water.

Marthas Vineyard Ponds

On the Islands, as we have seen, most of the ponds lie in beds begun by glaciers and completed by the sea. Many are the large ponds which fill onetime coastal inlets, natural low areas in the terminal moraine. Now walled off by sandbars, their former salt waters have drained back to sea by natural subsurface groundwater flow, and freshwaters have replaced them. A 1780 map of this coast shows Squibnocket Pond connected to the Sound even that recently by a channel wide enough for boat passage.

Squibnocket Pond, in Gay Head and Chilmark, is only one of the several that line the Vineyard's northern coast: Farm Pond in Oak Bluffs; Lake Tashmoo in Tisbury; James and Daggetts Ponds in West Tisbury; Squibnocket; and Poucha Pond on Chappaquiddick. Sengekontacket Pond in Oak Bluffs would be the largest of all, but it has not quite torn itself away from the sea yet; its very name means "salty waters." Some day it might complete the 500-foot wide bar along which the Edgartown road now runs. The bar is kept open artificially now for the sake of shellfish which live in Sengekontacket's sheltered waters.

Across the Vineyard, along the southern shore, a parallel series of long, narrow ponds lines the outwash plain, as we have seen. These too are sand-dammed inlets. They lie in grooves which once were channels carved by glacial meltwater streams. Then, as on the north of the island, the sea rose. It crept into the lowlands and lay there, until its own currents plastered bars across their mouths.

No kettle holes dimple this outwash plain, nor that of Nantucket. These regions lay beyond the farthest reach of any ice. A few blocks, however, apparently did split off along the front of the glacier as it began its retreat. The terminal moraine thickened about these blocks, and kettle holes remained sprinkled in the morainal areas of the Islands.

Kettle ponds on the Vineyard are common and lovely sights throughout the uplands. On the west there are Seths and Old House Ponds in West Tisbury; on the east, Fresh and Dodger Ponds in Oak Bluffs, and Lily and Jernegan Ponds in Edgartown. On the other hand, there are some ponds among the hills, such as Bliss Pond in Chilmark, which result from the damming of small brooks.

Nantucket Ponds

Nantucket, close cousin that it is, has a pond arrangement wholly similar to that of the Vineyard. One of its largest, Sesachacha Pond, is a natural break in the eastern end of the moraine. At one time, doubtless, it was a harbor similar to Polpis Harbor, but a sand barrier has severed it from the sea. Copaum Pond, northeast of Trots Hills, is similar. As for the southwest coast, the outwash plain is furrowed by a series of long ponds very much like those on the Vineyard. Fishermen have found the sandbars at their mouths quite handy. At high tide, fish swim over the bar into the pond. The tide wanes, and out pour the fish through narrow gateways into the fishermen's nets.

Among the hummocks of Nantucket's low moraine there are kettles. In the western hills lie North Head Long, Washing, Maxey, and Head of Hummock Ponds east of Trots Hills. Saul's Hills contain Gibbs Pond and dozens of smaller ones. Swamping over of ponds is so common on the island that everywhere along the moraine are blue dots with wide borders of bright green, indicating the original size of the ponds. Such are Reed and No Bottom Ponds west of Nantucket Town. And throughout the moraine, most of the swamps themselves are all that survive of former ponds.

Cape Cod Ponds

"By a beneficent arrangement of Providence," said an early Cape historian, "these ponds, containing an article so necessary to life, are found in almost every part of the Cape." Indeed, Cape Cod has more ponds than it has square miles of land. The writer has counted 466 on United States Geological Survey maps. In a view of Cape Cod from an airplane, we see so many ponds riddling the land surface that we are likely to wonder how some parts of the Cape hold themselves together from shore to shore.

Although man has seen fit to add still more ponds, most of those on the Cape are natural. They range in size from tiny pools less than an acre to lakes more than a mile long. By far the largest is Long Pond in Harwich-Brewster; it covers 743 acres and is sixty-six feet deep. Mashpee Pond is half as long and two feet deeper. Wequasset (or Chequasset) Lake in Barnstable, scalloped with coves and inlets, is 644 acres. This is the lovely pocket of blue which we see south of Route 6 where the road climbs to the top of the Sandwich Moraine in Barnstable (*Figure 25*). There are about a hundred other ponds with areas of twenty acres or more. Most of these, together with Vineyard ponds, are stocked by state fish hatcheries in Sandwich, with brook, brown and rainbow trout.

In total area, pond waters cover a little more than a third of Cape Cod. Their great number, as we have seen, arises from the fact that ice spanned the entire Cape. For kettles hold most Cape ponds. In addition, there are a few which lie in natural irregularities in the drift itself. Some of these near shore, as on the Islands, represent one time harbors, dammed now by sandbars and turned into fresh lakes. And still others, those found in more recent sand deposits such as those in Provincetown, lie in natural hollows among sand hills.

Sometimes a pond, by a sort of asexual fission like the amoeba's, becomes two or more ponds. Narrow bars maintain the separation. In Chatham, west of Great Hill, Old Corners Road crosses a bar separating Stillwater Pond from Lovers Lake. Long Pond in Wellfleet nearly has pinched itself apart in the middle. What once was Gull Pond, also in Wellfleet, is another example; it has become Williams and Higgins Ponds (*Figure 36*). The waters of these, however, still mingle via low drainage channels through the bar. A bar across Wequasset Lake in Barnstable is nearly complete (*Figure 25*).

Such a bar is the pond's own creation. Storms can turn the quiet surfaces of lakes into miniature seas of wind-driven waves and surf. Just as huge breakers of the open ocean rake into morainal cliffs of the shore, so little breakers of the lake attack the drift hills of their shores. And as the sea takes its plunder and with it builds bars and spits, so in miniature do the lake currents. Wherever the lake happens to be narrow, where land pinches in on both sides, these land bulges deflect currents, causing them to drop their burdens. In time a bar forms. Eventually this may reach all the way across the narrowing strait in the lake, cutting the lake in two. The separate ponds, like good neighbors, keep the fence repaired.

There are ponds too which are encircled by offspring, like hens and chicks. Such is Great Pond in northern Wellfleet, with Northeast, Southeast, Southwest and Turtle Ponds, one-time irregularities in the shoreline, surrounding it.

Rings of boulders at water level neatly outline many Cape-Island ponds. The rocks may be a foot or two across. So conspicuous may be these rock borders that in some places, local tradition ascribes them to work of Indians: "The pond shore was their camp site, their backyard, after all, and they laid rocks along it to pretty

Figure 36. *Gull Pond, Wellfleet, Cape Cod. This is one of the Cape's many kettle ponds.* NATIONAL PARK SERVICE

it up." Such rock rims, however, are common to lakes in all glaciated areas subject to cold winters. Perhaps the Indians did take time to give nature a hand in building them, but there is little doubt that nature did most of the work.

Ponds in glacial terrains have no lack of good-sized cobbles and boulders on their floors. Wintertimes, when surface waters turn to ice, sudden cold spells cause the ice to shrink and crack. Contracting, it may pull itself back from the shore, as any ice skater knows. The water thus exposed will freeze too. Because it has contracted, the ice becomes tightly packed in the pond. When the air warms, the ice must expand. With no place to spread, it pushes against the shore. The force of expansion is inexorable, and the crowding ice shoves before it debris from the pond bed, even large rocks. Thus, all along the lakeshore a rampart of lake-bed materials —stones, pebbles and sand—forms at the winter water level. When the season's ice melts, sand sifts away back down the sloping sides of the basin, but large materials remain.

Groundwater

Every freshwater pond of the Cape and Islands, as well as every brook and spring, owes its existence to the groundwater supply. Groundwater is the water which saturates the ground below a certain level, called the water table. It fills all available spaces in the rock or, in our Cape-Island case, in the till. It lies between sand grains, in pore spaces in solid rock, in tunnels and cracks caused by plants and animals, and in shrinkage cracks in clay.

The groundwater supply is sustained by rain and snow. Upon landing, only some of such atmospheric water runs off along the surface. Another part evaporates. The rest trickles down to the water table. Since the ground is saturated below this level, newly arrived water can go no farther down but lies above the water table as it finds it, creating a new, higher one locally. Thus the position of the water table varies somewhat from time to time and from place to place.

In a casual way the configuration of the water table follows that of the land above it. Where there is a hill the water table also rises, but not as steeply. Where there is a hollow, the water table swings down, but not as much. Where the land surface plunges suddenly, the water table, sloping less, may intersect the surface of the land and emerge as springs. These may feed swamps, ponds or streams. Where the land drops down to sea level, the water table too meets the sea.

Natural hollows in the land which dip below the water table become filled with groundwater to the height of that particular part of the water table where they occur. Thus, the level of every freshwater pond of the Cape and Islands represents the level of the water table at that point. That is why the heights above sea level of pond surfaces, regardless of the depths of the ponds, are lower the nearer the ponds are to the sea; for the sea level is the level to which the water table descends. However porous the sandy floor of a pond may be, unless a severe drought causes the water table to drop unusually far, the pond water will not seep out, for every pore of the sand below and around it already is saturated.

Swamps

Water-loving plants have taken over many of the shallower kettle holes on the Cape and Islands, and have transformed them to swamps. Other swamps line the borders of deeper ponds. In some areas the swamps are the last strongholds of undisturbed wildlife (*Figure 37*).

Figure 37. *A kettle pond at Brewster. Swamps begin when water-loving plants venture from shore.* B. CHAMBERLAIN

The first step on the pathway from shallow pond to swamp is the growth of pondweeds. In the sand and mud of the pond bottom these anchor themselves by the roots. When they die their remains build up a vegetable mold on the pond bed, and the level of the pond floor rises. Plants at home in shallow water can venture farther from shore. White water lilies, duckweed and spatterdocks send their tuber-like roots down, at the same time hoisting their lily-pad leaves to the water surface and spicing the air with their flowering fragrance. As these too die and decay, more vegetable mold builds up on the pond floor and it becomes increasingly shallow. Finally such actual swamp plants as cattails, pickerelweed and bulrushes invade the area and displace the water almost entirely. As long as the floor of this swamp remains at just about the water table level it will remain a swamp. When vegetation builds the floor up to above the water table, ordinary land plants take over and the swamp is gone.

On the Cape and Islands, swamps are common sights. Among the Cape's largest is that in South Dennis along Swan Pond Road. It covers a square mile. In Harwich, groups of swamps cover five times that much land.

The center of swamp and bog population on Marthas Vineyard is on the poorly drained highlands of Gay Head. On Nantucket, swamps have just about replaced all of the island's former ponds.

Many mainland swamps seem dreary and unfertile compared with those in this entire Cape-Island region. Here, drying winds and sandy soil discourage the growth of the usual spongy peat mosses. In their place spread bushes and trees—red maples and cedars—which fertilize the swamp soils with their rich remains.

Cranberry Bogs

Great are the riches which swamps have given to the Cape and Islands. They have been the framework of the multimillion-dollar cranberry industry. The cranberry, that tiny red evergreen fruit, saved Cape economy from almost certain collapse in the years immediately following the Civil War. And it still is extremely important today.

Wild cranberries are native to this area. Early in the nineteenth century it was discovered that these plants seemed to prefer swamplands and produced their best berries when the plants were covered with sand. In fact, on Gay Head, Marthas Vineyard, wild cranberries still grow conspicuously. The winds keep them amply supplied with a sandy blanket. The Indians there celebrate a yearly Cranberry Day in the autumn, and the entire town turns out to pick the berries.

On the Cape too, everything needed for good cranberry production is present in abundance. The swamps are particularly valuable because of the richness of their soils. During the forty years after the discovery of cranberry cultivation, it swept the Cape like a fad, but one of vital importance.

Cranberry bogs are common almost everywhere on Cape Cod. Their heaviest concentration is along the northern half of the Upper Cape. On Marthas Vineyard also, most bogs lie in low areas of the northern moraine, especially in the Tisbury region. Nantucket has the largest bog of all. Indeed, it is said to be the largest in the world, stretching more than a mile wide southeast of Gibbs Pond.

Actually, to say "cranberry *bog*" is to speak loosely. Fields of cranberry cultivation really are not bogs in the ordinary sense. It is not the wetness of swamps that the berry plants prefer, but the peaty soil. In fact, not every cranberry bog has a swamp in its past; some are prepared from low, flat areas with rich soil. In most cases, however, swamps have proven the best starting places.

To create a cranberry bog from a swamp, the first step is to clear and drain completely. Then ditches are dug across the fields in the familiar crisscrossing patterns. The plants are set in and about four inches of sand is thrown over them, to insulate and hold them in place. During winter, the bogs are flooded to prevent severe freezes by providing a frost-warding blanket of mist and ice over them. In some bogs, birdhouses are set on poles to encourage birds to live there and keep the insect population down. On the bog margins, beehives here and there attract bees which aid in pollination. By now we no longer could recognize the former swamp. But the green cranberry plants are reaching down beneath the sand, drawing their nourishment from the remains of the plants which created the swamp. In the autumn, when these patches of bog are wine-red with the ripened fruits, the Cape and Islands reap the harvest.

Peat

Cranberry profits are not the only dividends which Cape-Island swamps have paid, nor the only way they have rescued citizens in distress. The abundant maple and cedar swamps, during hard times, have supplied fuel in the form of peat, one of nature's storage batteries of solar energy. Peat is seldom used for fuel today. But during the bleak Civil War years, after the forests had been destroyed and before the saving pitch pines had become widespread, women harvested peat from swamp and from sea cliff. Yearly they dug vast quantities, blocking it into large bricks and drying it. Wartime made wood nearly impossible to import. This swamp peat, found in nearly all the towns, became a veritable lifesaver; for houses had little by way of heating systems except their several fireplaces. Although temperatures are relatively mild throughout Cape-Island winters, the winds are uncompromising. On Nantucket, the average February wind velocity is more than fifteen miles per hour. As for Cape Cod's winter winds, Joseph C. Lincoln recalled them vividly from his childhood: "The wind does blow on Cape Cod. . . . Out of the northeast or the northwest [the winds] come, over miles and miles of tossing, berg-dotted salt water, and when they really settle down to business, their piercing breath mocks at temperature records. 'Only fifteen above zero?' they seem to say. 'Not really cold at all? Is that so! Well, *we'll* show you!'" But close by their peat fires, houses banked with seaweed, people did not mind the sound.

Iron Ore

The hidden treasures of the swamps did not stop with peat. Some Island swamps played a part in the winning of the Revolutionary War and the War of 1812. Such is Iron Ore Swamp on the Vineyard's North Road. As we can judge from its name, this swamp yielded iron ore, and a considerable amount. The ore was low-grade, spongy bog iron ore. This is a mixture of peat and a yellow-brown, water-rich iron oxide mineral, limonite. Only about one-quarter of the bulk of such ore is iron, but the fighting Americans could ill afford to be choosy. It was cheap to mine and the iron foundries of Taunton and Carver on the mainland willingly accepted it for their cannonball manufacture.

Such ore is particularly common in swamps of glaciated regions, where the dark-colored iron-bearing minerals in glacial drift supply the iron. Through postglacial centuries, percolating rain and groundwaters remove the iron from the drift and carry it along as they move through the ground. When the groundwater reaches a swamp, the activities of iron-precipitating one-celled plants, or other chemical reactions, may cause it to be redeposited.

There was at least one iron ore swamp on Nantucket. Similar ores were mined elsewhere throughout the New England-New York region during the early wars. Those from the Vineyard made about six dollars a ton. And they helped to make history. Cannonballs molded from them fed the guns of *Old Ironsides*—which, incidentally, was built by a Vineyard man, Chilmark's George Claghorn.

Streams

Concerning the streams on the Cape and Islands, Thoreau summed up the situation pretty well: "The least channel where water runs, or may run, is important, and is dignified with a name." For streams are far from conspicuous anywhere; most of those which exist have earned their title only by default. Some trickle from spring-fed ponds as they overflow more rapidly than evaporation can take place. Others begin as simple springs. Their water, augmented by other springs along the way, flows in channels previously carved to below today's water table by meltwater streams.

Nantucket Streams

Sand and gravel regions have even less than their fair share of streams. Rainwater is blotted up immediately and gets no chance to run on the surface. Such porous materials underlie nearly all of Nantucket. Here streams are small, weak and rare. The most impressive—if such is the correct word—are the brooks in the southeastern part of the island, feeding Hummock, Reedy and Miacomet Ponds. At one time there even were fulling mills along them. These brooks flow in oversized, ready-made channels. Most of Nantucket's other streams are confined to the northeastern part where, with a tenuous hold on existence, they wander through the swampy, poorly drained lowland.

Marthas Vineyard Streams

Unlike Nantucket, Marthas Vineyard is suited to support running water, at least in its western part. Much of the surface of its higher lands is underlain by preglacial and glacial clay. Here water gathers on the surface, for it cannot seep downward. Yielding to gravity's pull, it flows as streams from higher to lower lands. Black Brook at Gay Head is an example. It heads in a peat-filled swamp and flows, dark with peat, into Squibnocket Pond.

Black Brook, like all of the streams which rise in the clay-floored morainal upland, flows rapidly. But the streams of Tisbury and Chilmark are in the greatest hurry. Early inhabitants put all of these energetic brooks to work as millstreams, for the manufacture of many of life's necessities was once a strictly local affair. Mill Brook (the one in West Tisbury), which flows east and then south into Tisbury Great Pond, was dammed for a flour mill, creating Woods (Fisher) Pond in 1850, and Crocker and Priester Ponds farther east. At Crocker Pond we find a regular river system, for Mill Brook is joined by a tributary, Witch Brook. West of this is Paint Mill Brook, the mill of which ground local colored clays to make paint pigments. Farther southwest is Roaring Brook; this once powered a grain mill, a china clay works and a brick mill. In southern Chilmark, Fulling Mill Brook flows into Chilmark Pond; once cloth was cleaned here in the fulling mill. New Mill Brook, which rises as springs in the highlands and flows to Chilmark Pond, helped to grind grain.

In addition to these working streams there is the Tiasquam River, which rises in the moraine, flows east and north to West Tisbury, where it is dammed to form Davis and Looks Ponds, and arches back southward to flow into Tisbury Great Pond. All of the foregoing are morainal streams. There are others which use abandoned channels in the outwash plain; Mill Brook flows south from Priester Pond via one such channel.

Cape Cod Streams

Most of Cape Cod's fresh running waters, like the Vineyard's Mill Brook, have selected ready-made channels. They cluster in that part of the outwash plain between West Falmouth and Marstons Mills. So prominent in the landscape are the largest that they have merited the title of river, rather than the usual brook (*Figure 22*). There is Coonamessett River, heading in Coonamessett Pond; Childs River from Johns Pond; Mashpee River from Mashpee Pond; Cotuit River from Santuit Pond; Little River from Lovells Pond; and Marstons Mills River, heading in a swamp.

Elsewhere on the Cape, brooks and streams are rare and rivers nonexistent, except for the saltwater type. The largest body of fresh running water on the eastern half of the Cape is Brewster's Stony Brook (*Figure 26*). As we have seen, it rises in Brewster's Mill Ponds and, flowing rather rapidly down the northern face of the moraine, heads for Cape Cod Bay. From 1670 to 1870 there were five mills along Stony Brook, for grain, weaving and knitting. They earned for West Brewster, one of America's first factory towns, the name Factory Village (also known as Winslow's Mills)—a name which lingers still in tradition, although the factories are long since gone.

Herring Runs[15]

Early every spring, about the time the first mayflowers are opening in the woods and the throb of life is everywhere, manna falls on the Cape and Islands. Or rather it swims into them from the sea. This is when the alewives run. These close cousins of herring glint up the streams by the thousands upon thousands, their goal a lake in which to spawn. The streams become almost choked with alewives, struggling against the opposing water, jumping and climbing small waterfalls and mill sluices, driven by the most intense of

[15] For a fascinating and detailed account of a Brewster herring run, read *The Run* by John Hay (*see Bibliography II*).

urges. We could dip up netful after netful for hours and have no effect on the tide of fish. The stream edges snag the dead and dying, for the hazards of the trip are great. Thus, when the alewives run you can spot the locations of the streams from far away by the congregations of gulls which wheel overhead and dive for tidbits.

So important to early inhabitants was this springtime harvest of fish that millowners had to build their mills in such a way that they would not interfere with the movement of the alewives. Battles raged between milling and fishing interests. In 1806, for instance, the milldam which went up on Coonamessett River and blocked the path of the fish led to a long-term dispute with much bitterness on both sides. Another source of trouble was the building of the Cape Cod Canal. The earth banked up for it interfered with alewives which once had traveled up Monument River to Great Herring Pond north of the Canal. This was solved by a passage dug in 1917, which again opened this route to alewives and let them swim up the Canal.

In some places, artificial canals have been dug to obtain even more fish. These lead from the sea to ponds which otherwise would have no herring runs. One such is Herring Creek at eastern Gay Head. It follows a natural lowland and connects Menemsha and Squibnocket Ponds. Mattakeset Herring Brook in southern Edgartown lets herring into Edgartown Great Pond from Katama Bay.

The alewife industry is still important, not so much for food as for the manufacture of fertilizer. Some towns have herring rights. Residents may help themselves to a certain amount of the fish and must then pay to take more. One town on the Sound not long ago sold its herring right for 11,000 dollars.

Springs

The tremendous incidence of fresh surface waters on Cape Cod and the Islands points to an abundant freshwater supply below the surface as well. In most places we should not have to dig far to find it; in fact, it frequently breaks through the surface as springs.

If we wade a short distance out into Cape Cod Bay at low tide, the sun-warmed, soft oozing silt beneath our bare feet suddenly gives way to a patch of whitewashed sand, clear and ice-cold. This indicates a spring. The fresh water which trickles forth here continually washes away the mud, leaving only white sand. Here, we know, the water table meets the sea.

But why is fresh water concentrated in particular spots like that? In this case it is because most of the Cape where it borders Cape Cod Bay is morainal. There is impervious clay as well as porous sand. We may expect springs throughout such a region. Patches of clay lie in various positions within the till. In some places they may slant up to the surface and force out the groundwater which flows on top of the clay, unable to seep through it. The lay of the land may produce other springs. Where an irregular glacial terrain takes a sudden dip to below the water table, water trickles out.

In the uncrowded past, many surface springs were used for drinking water on the Cape and Islands. Today, the growing population with its ever greater water requirements has so lowered the water table that many springs have vanished.

As we have seen, this was the fate of the most famous Cape Cod spring of all, the Pilgrim Spring near the marsh behind High Head, North Truro. Gone too are the Cape "health" springs (for what is healthier than fresh water?). At the turn of the century, the "Cape Cod Pilgrim Mineral Spring Bottling Company" of South Wellfleet shipped out more than 15,000 gallons of springwater, which poured from one gravel bed at the rate of fifty-four gallons a minute. Yarmouth had springs too, which flowed nearly as abundantly. Apparently these did not awaken commercial aspirations.

Some springs still flow, of course. Some are safe for travelers to drink. At Gay Head above Herring Creek, safe water issues cold and sweet. This is truly a mineral spring, like many on the Vineyard. Organic and mineral remains saturate the Cretaceous and Tertiary sediments and show up in the springs as large amounts of iron and aluminum, and sometimes of hydrogen sulfide. Vineyard Haven, like some other towns, still taps some springs of the clayey morainal uplands. To the Indians, the Lake Tashmoo region was *Kuht'ashim'oo* (where there is a great spring). From that generous Tashmoo Spring at the head of the lake came Vineyard Haven's first town water system. The long-ago townspeople of Vineyard Haven, with their usual ability to put nature's gifts to good use, bottled the springwaters as mineral waters, and there is no question that they were just that. Lagoon Pond, east of Vineyard Haven, also has copious springs. Oak Bluffs still gets its water supply from

Wequi-tuckquoi-auke (place of the boundary spring). A bottling plant went up here too, from which "carbonated temperance beverages" were shipped out to whomever was interested.

Even Nantucket is well supplied with springs. For years the sand near Polpis Harbor has yielded running water. The water which blots so rapidly down through the island's loose sands joins into a vast sheet of groundwater flowing seaward. Sometimes glacial clays interrupt the subterranean flow and force water to spill out onto the land.

Water-Dowsing

Today on the Cape and Islands, town reservoirs take the place of springs in many towns. Outside of the towns, drinking water comes from individual wells. Which takes us to water-dowsing—of a scientific sort.

Other conditions being similar, a general rule worth remembering in water seeking is: wells drilled in depressions tend to reach water at less drilling depth than wells drilled from heights. In depressions, as we have seen, the land surface dips down more than the water table and they approach one another. On heights, the water table climbs less steeply than the land and is left somewhat behind. Another rule: on the Cape and Islands, the outwash plains and postglacial sand deposits tend to hold more water under their surfaces than the morainal belts. The principle is obvious; for a barrel of marbles would have more room for water in it than a barrel filled with a mixture of marbles and shot. Sand or gravel deposits, like the barrel of marbles alone, consist of relatively equal-sized grains. More water fits between the grains than between unequal-sized materials, where the small particles take up space between the larger. Till is a mixture of boulders, gravel and clay; it has less room for water than simple sand.

At the opposite extreme from outwash plains are regions almost wholly compact clay, such as western Marthas Vineyard. So greatly does the clay slow up water penetration that in some places an adequate supply is not always a certainty.

In any region, of course, the way in which the stuff of the earth itself affects water is modified by various trimmings. Plant roots in the upper soil zone provide openings through the ground for water to run through quickly. Engineering feats of animals and insects may reach down twenty feet or more and greatly increase porosity in certain spots.

Water-finding problems, all told, could hardly be considered severe on the Cape and Islands. On the mainland, many a home builder would rejoice to strike water at one hundred feet, but Cape Codders have been known to regard such a great depth as an unreasonable perversity of nature. Through their soft materials, drilling is usually quick, cheap and shallow. It is rather unusual for Cape-Island wells to go a hundred feet deep.

However, all is not always idyllic. Trouble may lurk in the hardpan clay or occasional boulders. Tricky perched water tables may lie beneath the surface. These are discontinuous patches of clay which prevent rainwater from seeping through locally to the true water table beneath the clay. Thus, a false water table forms above the clay. When tapped by a well, such a thin lens of water is likely to give out quickly during drought.

Another problem may be overcrowding. Heavy use of groundwater may reduce its level dangerously; Provincetown has had one of the more serious problems along these lines. In 1841 an observer wrote of the town: "Although near the ocean on every side, the inhabitants obtain good water by digging a moderate depth a few feet from the shore." In recent years, however, the water table has been forced downward into the shape of a cone of depression by too much pumping of water to serve the needs of the town. If the cone were allowed to go still deeper, seawater might seep into the wells. Pumping is therefore regulated by rationing it when the water table is low, thus maintaining its safe level.

Localized in their occurrence and nearly impossible to predict are Cape-Island artesian wells. Wherever an impervious clay bed lying just at the water table is struck by a drill, it is not likely to yield water; the drill must go through to the more porous material beneath. But the water in this porous material is compelled to seek its own level, the water table. Thus it must rise, if given the chance to do so, to as high as the clay. The passage made by a drill frees it and permits it to rise. This makes the well artesian, its waters rising above the level at which the drill finds them.

Perhaps the most wondrous water finds of all on the Cape and Islands are the freshwaters which lie not far below the surface of salt-sprayed sand dunes and beneath the sands of all the beaches.

Testifying to the presence of such water are the plants which thrive on the white sands.

When Henry Beston lived in his Outermost House among the Nauset dunes, he obtained water from the sand:

> The top of the mound I built on stands scarce twenty feet above high-water mark, and only thirty in from the great beach. . . . To get drinking water, I drove a well pipe directly down into the dune. Though the sea and the beach are alongside, and the marsh channels course daily to the west, there is fresh water here under the salty sand. This water varies in quality, some of it being brackish, some of it sweet and clear. To my great delight, I chanced upon a source which seems to me as good water as one may find anywhere.

Now the entire Nauset Beach strip from Provincetown to Chatham may yield up its fresh underground water to thirsty fishermen and beachcombers. Plans have been underway to place a series of hand pumps there under the auspices of a conservation club. Some of these wells will need to go only six feet deep to pierce the water table. When water is reached it must be pumped until clear, after which very pure drinking water is obtained.

On Nantucket too, dune waters serve as drinking water. Great Point has a series of low dunes. The lighthouse keeper gets his drinking water supply from a depression within these dunes. The water remains fresh even when storm waves intrude from the sea beyond the dunes. Frederick Pohle suggests that this seemingly miraculous freshwater supply may be the "dew" which thirsty Vikings found somewhere along this coast nearly a thousand years ago.

CHAPTER X

Salt upon Salt

The turtle-like sheds of the salt works were crowded into every nook in the hills, immediately behind the town [Provincetown], and their now idle wind mills lined the shore. It was worth the while to see by what coarse and simple chemistry this almost necessary of life is obtained, with the sun for journeyman, and a single apprentice to do the chores for a large establishment. It is a sort of tropical labor, pursued too in the sunniest season; more interesting than gold or diamond-washing, which, I fancy, it somewhat resembles at a distance. . . . It is said, that owing to the reflection of the sun from the sandhills, and there being absolutely no fresh [surface] water emptying into the harbor, the same number of superficial feet yields more salt here than in any other part of the country.

HENRY THOREAU
Cape Cod

Not only the surface streams, but the waters of most of the vast natural water reservoirs beneath Cape Cod and the Islands ultimately reach the sea. To the sea they contribute salts, dissolved materials that they have picked up on their land travels. And elsewhere, all over the world, other groundwaters and other rivers also feed the sea. They pour into it billions of tons of the elements dissolved from the rocks of the land, and many more billions of tons of solid alluvium. The dissolved materials, carried by currents and scattered through the waters by diffusion, spread to every corner of the sea and keep it salty. The sea is a huge storage vat for the soluble materials leached from the earth.

Go where we may on the Cape and Islands, we cannot get beyond the tang of the sea's salt winds. But it is along the beaches especially that the sea endlessly returns the elements received from the land—in temporary, Indian-giver fashion.

On the beach at low tide we find a boulder within reach of high-tide waters, one with a shallow depression in its top surface. There are countless such rocks on breakwaters and jetties. The hollow in the rock sparkles with a white crystalline coating. We taste a bit of this and find it to be salt. Let us look at it with a magnifying glass. We see crystals, small cubic blocks piled crazily together. Their existence is transitory, for the water of the next high tide will break them down and once more mingle them with its other dissolved salts.

Such salt crystals form, of course, as a result of natural solar evaporation of salt water. When the warmth of the sun drives water from the rock surface, remaining water becomes oversaturated with salt and cannot hold it in solution. The salt takes its leave of the water by crystallizing.

The Sea-Salt Industry

"Salt upon salt may assuredly be made," reported Captain John Smith in 1616, when he noted the vast supplies of fish surrounding the Cape and the easy manner of catching and preserving them. Later, it was natural for fishermen of the Cape and Islands to look to the sea for the salt which they needed to preserve their catches of fish. The real impetus for a sea-salt industry was the British blockade during the Revolutionary War, which cut off the supply of foreign salts. Salt fish was a basic industry, and a salt supply imperative. So Cape Codders and Islanders were forced to look to the limitless salt mines in their own backyards.

At first it seemed too slow and cumbersome a process to let the sun's energy do the work of evaporation. Instead, great kettles were filled with seawater. These were set over open wood fires until the water entirely boiled away and a coating of salt crystals lined the bottoms of the kettles. As the forests and their fuel supplies grew scarce, a Dennis man, John Sears, tried ordinary solar evaporation in 1776. Through trial and error and amid definite general skepticism, which dubbed the project "Sears' Folly," he perfected his method. Within a score of years, towns all over the Cape and Islands had cast off their doubts and were producing salt.

Generally, the process was to lay out long wooden troughs, about thirty-six by eighteen feet, and less than a foot deep. Windmills pumped seawater into them. The water was allowed to evaporate in stages in the sun. Ever-brinier water was pumped from vat to vat to extract different salts. In case of rain there was a shutter arrangement to close off the troughs. It took a long time—up to six weeks—to get a respectable amount of salt, but with the scarcity prices of wartime the process was initially profitable, and once established, it continued to pay off.

The most abundant salt in the sea is halite, or sodium chloride, or common table salt. Thus, halite was the big product in the Sears process. But even though halite makes up nearly four-fifths of seawater's dissolved mineral matter, the water first must evaporate until only one-tenth of its original volume remains before it can no longer keep halite dissolved and the salt comes out. Still more water must go off to leave the still more soluble and less abundant salts of ocean water. These include magnesium sulfate (epsom salts), sodium sulfate (Glauber's salt), magnesium chloride and potassium chloride. Sometimes calcium sulfate, less soluble than halite, may precipitate out in advance of halite.

When a series of Cape-Island salt vats had lost all their water, the salts clung to the bottoms and sides in layers. These layers, from vat to vat, represented salts of increasing solubilities under those conditions of evaporation. The salts could be scraped out, the water pumped back to sea, and the vats refilled with new seawater.

The scraping of salts from the vats each time evaporation was completed was measured in inches. It took 350 gallons of water to obtain one bushel of salt. It would have taken tremendous volumes of water to produce sizable thicknesses of salt at one time. If a column of seawater the size of the Empire State Building were completely evaporated, the total thickness of salt left would fill only one and a half of its 102 floors. Of this, a little over one story would be table salt, about four feet would be all of the potassium and magnesium salts together, and eight inches, calcium sulfate.

Despite the sluggishness of the process, so many saltworks sprang up on the Cape and Islands that in 1802, the Cape's 136 works produced 40,000 bushels of table salt and 182,000 bushels of Glauber's salt. Tanneries used Glauber's salt to keep drying hides soft and supple. A sideline product was epsom salts, which came out during the winter, when the brine was first heated, then suddenly chilled.

Twenty years later, the saltworks had more than tripled and production had increased more than tenfold. In addition to the works at Dennis, Provincetown had sixty-nine; Eastham, fifty-four; South Yarmouth, fifty-two; Chatham, eighty. By the time of the Civil War, 500 had burgeoned from Sears' Folly. Travelers described the entire Truro and Provincetown shores as lined with vats and the creaking, clacking arms of the pumping mills. The Vineyard also had extensive works on the shores, particularly along Vineyard Haven's Lagoon Pond. Nantucket made an attempt and built large works at Brant and Quaise Points. However, there the industry did not last long, for heavy fogs which sometimes drift over the island hindered evaporation.

Begun from necessity as a sideline to the fisheries, the salt industry thus became vitally important in its own right. The two cannon which stand today on the lawn of the County Courthouse in Barnstable, Cape Cod, attest to this importance; they were brought from Boston during the War of 1812 and set up on the Cape shore to protect the saltworks.

By the end of the nineteenth century, increasing competition with salts from Europe and the American West became too much for the Cape-Vineyard industry. Despite a great production capacity, finally it was forced to close down. The vats were disassembled, the briny wood thriftly used for homes, and the arms of the windmills stopped turning forever.

Other Ocean Products

It is interesting to speculate why the most common dissolved salts of the sea are salts of sodium and chlorine and magnesium, rather than compounds of calcium and silicon. For rivers and groundwaters which bring these dissolved substances to the ocean gather them mainly from the rocks of the land. In the transfer of these waters from land to sea, subtle changes take place in their composition.

Rocks contain, in general, far more calcium and silicon within their mineral constituents than sodium and chlorine. Rivers reflect this and carry relatively large proportions of calcium and silicon in solution. But when they reach the sea, the balance tilts the other way. Seawaters contain far more sodium and chlorine than calcium and silicon. This, apparently, is the work of small sea creatures. Scorning sodium and chlorine, they seize and hoard calcium and silicon.

Calcium, for example, becomes an intrinsic part of the limy, hard parts of crabs, lobsters, molluscs, barnacles and corals, and of a vast number of one-celled animals such as the chalk-forming Globigerina. Usually this calcium, thus tied up, remains long lost to the water of the sea. Some creatures such as corals build reefs and eventually create lands with it. Others die and leave their shells to accumulate on the sea floor. Silicon is another popular commodity among sea animals. Some organisms use it for their hard parts. Sponges make skeletons of it. The extremely abundant one-celled sea plants, diatoms, form shells for themselves of silica in the form of opal, such as we see in the Gardiners Clay.

Salt no longer is an economical thing for Cape Cod and the Islands to produce. But perhaps someday they will turn back to the ocean and seek from it its more elusive mineral resources, as they have long combed it for its animal and plant harvest. Although the sea's inorganic resources are diffuse and dilute, they are in limitless supply. Each year the rivers bring about three billion tons of dissolved materials in addition to all the sediment. When you swallow a mouthful of seawater you are swallowing more than fifty elements. You are increasing your monetary value by nearly a billionth of a cent, the worth of the gold you swallow. For in addition to the very abundant sodium and chlorine, these other fifty-plus elements, including gold, drift about in solution in such gross amounts that every cubic mile of ocean water has more than a hundred and fifty million tons of them. Some, such as manganese and phosphorus, have precipitated into vast nodular deposits on the sea floor, similar to the phosphate lumps in the Vineyard greensand.

The less common elements in seawater, particularly such metals as magnesium, aluminum, lithium, manganese, copper, zinc, uranium and nickel, may someday prove worth going after. If they do, Cape Codders and Islanders again might mine the sea. For while men have searched long and wide and hard for ores of these metals, Cape Codders and Islanders have been hearing them daily, pounding at their own front door.

CHAPTER XI

These Prairies by the Sea

Yet these swamps are on many accounts the most remarkable features of the ocean shoreline. They probably represent a larger share of solar energy applied through organic life to constructive geological work than is shown by the coral reefs; this is certainly the case if they play the same part along all the northern seashores that they have along the coast of New England. . . . There is nothing in the coral islands to compare with the brilliancy of coloring which the marshes of New England exhibit at certain seasons of the year; nothing like the variety of hue which characterizes these prairies by the sea in the different seasons. . . . Nothing in the geographic world is more graceful than the curves of the creeks through which the tidal waters enter and depart in their constant movement.

NATHANIEL SHALER
Geologist, 1886, 1893

Like a vast patterned carpet at the foot of Cape Cod's Sandwich Moraine stretch the Great Marshes of Sandwich and Barnstable (*Figure 25*). More than 4000 acres of marsh grass ripple in the salt winds like inland fields of grain. Lush and green in summer, in the autumn these meadows put on hues ranging from maroon to "as tawny as the lion's skin," in the words of J. M. Mackie in 1864.

There are other salt marshes as well on the Cape and Islands. Ribbons of marsh wind up salt-creek channels and line the coves along all of the shores. Long irregular strips of marsh lie back of the Outer Beach of Chatham, Orleans and Eastham (*Figure 38*). Wide, flat marshlands cut into the Cape Cod Bay shore at Dennis and Wellfleet, and border Provincetown Harbor. On the Islands marshes dip in and out of the land with every wrinkle of its outline.

Nearly everywhere the marshes still are growing. Some have

reached a comfortable middle-aged equilibrium. Others have become severed by sandbar fences from the sea which gave them their beings, and have turned into freshwater swamps.

Values of the Salt Marshes

Cape Cod and the Islands are twice blessed in their marshes. Theirs are the most extensive stretches of salt marsh in Massachusetts. Theirs also are among the most unspoiled marshes of the entire eastern United States, and here the Cape-Island region is especially fortunate. It is easy to destroy a salt marsh. Along much of America's coast, carelessness, self-interest and ignorance have helped to eradicate the marshlands and their loveliness—forever, as far as we and many generations to come are concerned. Bulldozers tear into them. They are turned into junkyards and garbage dumps, or drained for shorefront housing developments. This vandalism—for to such it often amounts—involves more than the creation of something ugly from something uniquely beautiful. It involves a serious disturbance in bird populations that breed in the marshes. And it may well mean the eradication of commercial fisheries, for the delicately balanced estuarine environment of marshes is the nursery of many important fish species.

Marsh Environments

Along shorelines, wherever barrier beaches develop or tidal flats indicate shallow water, marine marshes follow. They grow in the quiet heads of coves and spread themselves along the shores of saltwater creeks. Where shores are slowly invaded by rising seas, as the Cape-Island region has been, salt marshes replace drowned freshwater swamps and ponds. From small beginnings the marshes spread, stealing a little from the land and a little from the sea, channeled by delicately laced and intricately changing creek beds, and meadowed with grasses.

The Great Marshes of Sandwich and Barnstable started to develop when the irregular ground moraine shoreline of the Bay became barred by sandspits from Sandwich to Barnstable. These necks of sand, built probably around cores of glacial drift, are major additions to the land. Town Neck, Springhill Beach, Scorton Neck, and large Sandy Neck all trail eastward. Sandy Neck gives Barnstable its harbor.

The marshes behind Nauset Beach sprouted forth under the protection given them by long, narrow barrier beaches (*Figure 38*). At Chatham, opposite the Chatham Bars Inn, the southern part of Nauset Beach shelters low, marshy Tern Island from the sea. Here, nature's sandbar and man's Audubon Society protectively allow 8000 pairs of terns to live and breed in safety. From here each year they fly to Europe, the Falkland Islands and South America, then back again to Nauset to the same spot from which they started. Cape Cod is a way station on one of the four great bird migration flyways of North America, reaching from the Arctic to South America. Its marshes, coves and beaches such as those of Tern Island serve as resting places for a wide variety of bird species, as well as nesting places for some. Diligent bird watchers

Figure 38. *Growing marshes. These marshes at Nauset are protected from the sea by sand dunes (background), and are spreading. In the foreground, the edge of the kame fields.* NATIONAL PARK SERVICE

can spot some 250 species among the tens of thousands of individuals on the Cape in a single year.

South of Tern Island is another marsh and sand specialty island, Morris Island, site of a rare example of a white cedar swamp.

At Wellfleet on Cape Cod Bay, a large marsh of nearly 3500 acres has closed an inner harbor in which ships once found anchorage. Here bars have risen and bridge the lowlands between Bound Brook, Griffin and Great Islands and Great Beach Hill. These bars form protection for the large marsh tracts behind and between the "islands," which sprawl where tidal creeks once flowed. Before the sandy connecting bars had developed, storm winds sometimes must have driven the sea into these lowlands. As water today planes flat the bayside cliffs north of South Truro, so it must have had access to the morainal ridges and hills which surround the Wellfleet marshes; for today these hills are sharply truncated by similar steep cliffs, even though waves no longer beat against them.

Thus it is with Chatham's marshes too. These spread out along the shores of Oyster Creek and Oyster and Mill Ponds, and around Morris Island. In some places they meet sharp cliffs, the results of that early premarsh period of wave erosion.

Tidal Flats

Wherever quiet, shallow waters lap Cape-Island coastlines, we are likely to find the familiar current-swayed blades of eelgrass. As we walk across the low-tide flats of Cape Cod Bay we see patches of it here and there, which reach as far from shore as we can go. Many a swimmer has felt it tickle his stomach at high tide. Parts of the Bay floor are green with it, like oases in the vast areas of bare sand. This perennial flowering plant is the foundation of the life structure of a marine marsh.

Eelgrass can grow anywhere from twelve feet below low-tide level to nearly the high-tide line. Its survival depends on submersion in salt water at least half of the time. Thus its growth range is essentially the offshore zone visible at low tide. Vertically, this is at most a few feet, but horizontally, where the flats slope gently, this growth range covers vast areas.

As eelgrass grows, its roots anchor the sand and mud of the flats and create a humus-rich, boggy deposit. In relatively open waters such as Cape Cod Bay, marshes get no chance to develop further. Rapid, undisturbed accumulation of silt and humus would be required if marsh plants other than eelgrass were to gain a roothold. But in the open Bay waters, muddy debris which unceasingly falls to the floor is just as unceasingly kept dancing by currents. Most of the Bay floor remains in the tidal-flat stage near shore (*Figure* 39).

The sand and mud of Cape Cod Bay's spectacular tidal flats have drifted in from the cliffs on the other side of the Cape along the ocean. Currents sweep these assorted materials around the tip at Provincetown and drop much of the sandier part onto the growing spits at Race and Long Points. Smaller sand grains, mud and silt, drift farther and finally settle to the floor of the Bay. The flats between Long Point and the west end of Provincetown are

Figure 39. *Evening on the tidal flats. The vast tidal flats of Cape Cod Bay stretch nearly two miles from shore in some places.* B. CHAMBERLAIN

widening slowly into the Harbor. Thus, the oceanside's loss is the bayside's gain.

"Before the land rose out of the ocean and became *dry* land," wrote Thoreau, "chaos reigned; and between high and low water mark, where she is partially disrobed and rising, a sort of chaos reigns still, which only anomalous creatures can inhabit." This bizarre landscape in miniature, with its welter of strange creatures, can be the most fascinating of places over which to go roaming. The low-tide floor of Cape Cod Bay is certainly the patriarch of all, but some of the smaller lagoons and bays are equally interesting.

Life and Patterns of the Flats

Evidences of the life which makes these flats its home are all around. The more carefully we look, the more we shall see. For instance, many of the shelled species of creatures found as fossils in the pre-Pleistocene sands of the Islands are living on the flats today in the same way and under much the same local conditions as their venerable ancestors. We see wood channeled with tubes made by a close cousin to the long-vanished molluscs that did the job on some of the wood fragments of the Cretaceous lignites at Gay Head. Perhaps most interesting of all, we may see a true living fossil plodding along, one whose ancestry takes him back some four hundred million years—past the glacial ages, past the Cretaceous, and back into time again as far, to that strange earth where tropical corals grew in Greenland and all life was confined to the sea. This, the clumsy, armored horseshoe crab (or king crab), has foraged since then almost unchanged upon the mud flats. More closely related to spiders than to crabs, it can be found only on the Atlantic coast of North and Central America, with a related species on the eastern coast of Asia—nowhere else. Having survived so long, it now faces extinction by the hand of man. Countless millions of tons are destroyed for fertilizer and pig feed each year. Bathers, in ignorance, unjustified fear, or sheer wanton slaughter, increasingly are killing hundreds on Cape beaches every summer. Surely such an interesting and unusual creature—whose direct ancestors shared the cradle of creation with creatures extinct these four hundred million years—should not be ruthlessly destroyed.

The interbalanced life of the tidal flats would require many volumes of its own. But the shape and stuff of the flats themselves are interesting too, and take us back to the realm of geology. Many and varied and sometimes beautiful are the patterns etched by all comers into these yielding sands. We cannot walk far in bare feet, for instance, without feeling the ripple marks which ridge the surface of the sand. These reflect the eddies and swirls of water set up by waves and currents in shallows on sandy bottoms.

Ripple marks etched by waves on the sand (oscillation ripples) differ from those produced by currents. When a wave passes over the water surface, it transmits a water movement downward.[16] On the shallow sea floor the water motion reaches bottom, where it takes the shape of a flattened ellipse, water swinging back and forth. Where there are small objects on the sea floor the water piles the sand about them into peaks and hollows like those on the surface of a meringue pie. Generally the peaks are no more than a fraction of an inch high and a few inches apart.

Currents, on the other hand, create ripple ridges much as the wind creates dunes, piling the sand in asymmetrical hills with the steep sides downcurrent, on the lee. They can be many feet from crest to crest and several feet high.

On the sand of the tidal flats, the pattern of ripples that we see more often than not is a compound one, a complex honeycomb combining both sorts of ripples. If, through some geological change, the upper sand layer of the flats were to become raised out of reach of the sea and hardened to rock, these marks might remain on the rock surface, like the ripple marks which have been found on the semiconsolidated Sankaty Sand.

And there are other patterns in the sand. Here and there on the flats, moon-shell snails may have left sinewy trails as they crawled along. Little periwinkles may have added their thread-like paths. There are footprints of scavenging gulls and terns, and the tiny tracks of hermit crabs. Holes of soft-shelled clams tattoo parts of the sand and may spout forth miniature geysers as we walk. Other holes may be tube openings in which sea worms live. Toward high-tide line there are river systems on a miniature scale, complete with tributaries, where rills and rivulets of leftover tidal waters channel their way back to sea. The branching shape sometimes is complex enough to form a diamond-like pattern across the sand,

[16] This is discussed more fully on p. 233.

and arises from tiny obstructions in the course of the water trickles—stones or bits of shell, or the feelers of sand crabs which lie buried beneath the sand, antennae up to trap food from the passing water. Finally there are fragile bumps under our feet, small sand balloons of trapped air pockets which deflate when we step on them and break their surfaces.

Future Oil Fields

Beneath it all, in many places, is a black layer of muck just under the surface silt. As our toes ooze down through this muck and disturb it, we smell traces of hydrogen sulfide gas, the perfumer of rotten eggs. This black muck and its gas are the products of organic decay of plant and animal matter; and this is the sort of place in which many of the petroleum-bearing rocks which supply the world with oil may have originated. In fact, so rich are the bay floors with organic matter that in 1940, geologist J. L. Hough tested the black mud beneath Buzzards Bay to see if, sometime long, long hence, it would be capable of becoming a source rock for oil. One cubic foot of this sediment, he found, could supply seven-hundredths of a cubic foot of oil. Buzzards Bay covers a hundred square miles and its black sediments are ten feet thick. Thus, this region theoretically could someday be capable of producing some 1,960,000,000 cubic feet of oil—which is some three hundred and fifty million barrels.

Eelgrass and Shellfish

The eelgrass which roots itself in these Cape-Island embryonic oil fields helps to provide shelter for many of the molluscs which make their homes there. Thirty-five years ago we could not have walked across this same region without tramping through acre after acre of eelgrass. This was true for all bays and inlets of the Cape and Islands where there was protection from the open sea. With eelgrass at its luxuriant best, such semiprotected waters could harbor thriving populations of sea life. Among them were scallops, oysters, quahogs, and other types of clam. In fact an entire pyramid of sea and bird life is based on a foundation of eelgrass. Scallops particularly depend on this plant. They eat the organic matter which clings to its leaves and drops to the sea floor. In addition, the eelgrass breaks the force of sediment-carrying waves which otherwise would reach into the bay and bury the scallops. When eelgrass throve, small, tender Cape scallops were very abundant, particularly on the Cape's Buzzards Bay shores and the Sound shores, where the water is protected and the tidal range not too great.

But in 1931 the eelgrass was annihilated. Like a blackboard swept by an eraser, the flats of this region lost more than ninety percent of the grass. Biologists did detective work and identified the culprit, a parasitic fungus grown virulent. Its ravages were catastrophic to the shellfish industry, for they were echoed by the disappearance of the shellfish, with a partial exception of clams. By 1934, four-fifths of the cod, molluscs and crabs of these waters were gone, and even numbers of such shorebirds as geese declined.

Fortunately, an isolated nucleus of healthy plants survived, quarantined in the clear waters of Pleasant Bay, Chatham. From here, the eelgrass slowly is finding its way back everywhere. Either the disease has disappeared or the plant has built up an immunity to it.

The return of the shellfish is slower. It took until 1947 to get a good-sized small-scallop population in Pleasant Bay again; another year to get small clams; and until 1949 to get grown-up scallops and the return of periwinkles, crabs and waterfowl.

The scallop industry, although it is now reviving the eelgrass and scallop fields of the Sounds, has turned for the most part to the deeper water sea scallops. At Orleans' Rock Harbor on Cape Cod (*Figure 40*), for instance, boats go out along the tree-marked channel to harvest these molluscs. At low tide we can walk along this channel and pick up limitless numbers of their large, iridescent shells—and go home and serve baked scallops in them. Sea scallops are a big industry for Chatham, Dennis, Yarmouth and Monument Beach too. But the little bay scallops are still rather scarce on the Cape; although on Marthas Vineyard, one of the island's most lucrative fishing industries centers in the bay scallops of Lagoon Pond.

As for oysters, the statement, "no part of the world has better oysters than the harbor of Wellfleet," was a rarely disputed contention during the eighteenth century. These shellfish had long

Figure 40. *Rock Harbor, Orleans, Cape Cod. Also shown is Rock Creek, a tidal creek.* CAPE COD PHOTOS

been native to Cape Cod flats; Indian shell heaps contain thick piles of their shells. Early settlers mined these piles to obtain shells for mortar, Cape-Cod-style. Champlain found so many oysters in what was probably Barnstable Harbor that he named it Oyster Harbor. They were abundant along the Sound shores too. Oysters, in fact, were the enticement that first caused settlers to find their way across the great outwash plain to southern Cape shores. Pocasset and the quiet coves of the Islands likewise produced many.

Since 1775, however, some disease which hit the oysters has removed them, and never again have they regained their onetime vigor. Now they will not even spawn in the same place they grew up, and must be pampered throughout their lives. Today, when great oyster shells are picked up occasionally on Barnstable's mud flats by beachcombers, they are empty.

During the last century, oysters brought from the south were planted off Wellfleet, where they waxed big and fat, and obtained "the proper relish of Billingsgate" (a shoal in Wellfleet Harbor). Today, oysters are seeded in such places as the still waters of Marthas Vineyard, famed as first-rate oyster nurseries, and there left to grow on seashells as spat. Then the shells with the adolescent spat upon them are gathered and transplanted to the sands off Cotuit, Osterville, Chatham and Orleans, whence they finally are harvested.

Clams are rugged individuals, and seem not to have cared much about the state of health of the eelgrass, although their numbers did decline a little when the disease hit. Mainly three types of clam inhabit the flats—long-necks or steamers; razor clams; and quahogs. Of these, the large, thick-shelled quahog, which is the cherrystone when small, is far and away the favorite on the Cape and Islands. It is happiest on muddy bottoms where currents provide good water circulation. Cape Cod Bay is ideal. The Cape's southern shore is too exposed. So are the outer shores of the Islands. In these regions, quahogs settle down only in the smaller bays and harbors. Occasionally seed quahogs, dredged from the deeper waters of these bays, are planted in shallower, more accessible areas.

Today, a relatively sparse supply of clams remains. Their non-commercial taking by the public is rigidly controlled by many towns, often with special clam wardens. Apparently the fate of clams is tied not so much to the amount of eelgrass, as to the numbers of people fond of quahog chowder.

Marsh Growth and Grasses

In fairly open waters, marsh growth stops at tidal-flat stage with eelgrass and shellfish. In more quiet protected areas, however, the eelgrass zone, if undisturbed, is bordered by and may be replaced by a well-defined plant succession. For example, in Nauset Harbor during the last century, only forty years were required for wide, barren mud flats to turn into salt marsh yielding some 300 tons of salt hay.

In such enclosed shallow bays and inlets, mud and silt arrive constantly. They settle through the quiet water to the bottom, making the water still shallower. It is during this stage that eelgrass thrives.

But its very thriving is responsible for its downfall. When the tide lies halfway out, several feet of water cover most of the eelgrass plants. Water therefore passes freely over the tops of the blades, and suspended mud and silt can find their way into the quiet nearshore water. And when the tide turns, the closely spaced stems of the grass break the action of deeper currents, so that departing sediment, dragged out with the water, also is caught and dropped. Day after day, year after year, the process goes on. In time the bay floor rises slowly, fed by every pulse of the tide. Finally the flats may be built up to above low-tide level. Then eelgrass no longer gets enough water to nourish it, and it yields to other plants.

During the eelgrass period, fringing the eelgrass zone toward shore is the domain of a tall, green, coarsely bladed beach grass called salt thatch. All it desires is daily salt wetting; it does not need as much irrigation as eelgrass. Usually salt thatch occurs in narrow zones along tidal creeks or the seaward edges of established marshes. If we walk across a belt of salt thatch, especially in summer, we have rough going. Shooting up in April or May, salt thatch can grow a yard or two high, with as many as 600 stalks per square yard. It makes the soil underfoot wet, spongy and generally unpleasant.

Shoreward of the salt thatch comes a zone of fox grass and slender spike-grass, at the fringes of high-tide limit. Fox grass grows in dense tufts and is somewhat easier to walk over, for its soil is drier and more springy than that of salt thatch. Above this, farther up on the beach, is the sandy region of late-summer-flowering salt goldenrod, together with black grass and other typical reeds and grasses of beach and dune.

Like eelgrass, all of these plants help to build up the level of the land by the humus they form. Since sea level currently is rising and seawaters slowly are creeping up Cape-Island shores, many of these grasses keep their places by continually growing in pace with the sea-level rise. We can find their roots reaching as much as ten feet down to where they started when the water was lower.

Marsh growth is a slow process. It requires decades of unchanging conditions. For an entire human lifetime, the only indication of marsh development might be wide mud flats and eelgrass stands offshore. Even where zones of other grasses have developed along the shore, the plants carry with them a check on their own expansion. The debris which they trap forms a well-defined platform which grows in area and height until it reaches high-tide level. Here it can grow no more. Here, consequently, the marsh will have spread as far as it can. Growth may stop even before this, of course. If the marsh has filled in a sheltered neck of a bay or cove, it will stop spreading when it reaches open water where winds and currents prevent the requisite mud from feeding it.

To add to the unmanicured beauty of the marshes there are the estuaries and salt creeks which wind through them (*Figure 40*). Irregularities in the marsh surface provide channels. Before long, some of these become filled by fingers of the running tide. Existence for these salt creeks is a constant battle against the ever-extending marshes which squeeze them into limited boundaries. In this struggle, the creeks must keep their channels open by eating into the less resistant parts of their banks. Rivers of the land flow in one direction only, but tidal creeks flow two ways. On their return to sea, according to Shaler, their pattern of channel widening may become the reflection of the pattern carved by the incoming tide, producing curves of near-perfect symmetry.

Sandwich was the first town to be settled on Cape Cod (1637). It was settled because of its vast marshes. This apparently limitless supply of salt hay drew some of the Pilgrims back to the Cape from Plymouth, for they badly needed fodder for their animals. As more and more of the Cape became settled, the 20,000 acres of marsh across its surface became known collectively as the "hay grounds." The main grass used in this respect was fox grass, for it was valuable not only as cattle food but also for roof thatching. Marsh hay was a major product of the Vineyard too, especially the black grass which grows in the upland region around such salt ponds as Chilmark Pond. This was dried to become winter feed for sheep.

When salt marsh harvesttime came around, men and horses had to don great oversized shoes to keep from being mired in the soggy ground. In some parts of the world today, salt hay still is harvested this way. The Cape-Island harvests were good. In 1904 the small village of Provincetown alone was yielding more than 200 tons of salt hay each year.

Mosquito Control

If we pick our way through a salt marsh some warm summer evening, we may wonder how the Pilgrims ever got their haying done. Salt marsh mosquitoes are large, loud and voracious. But they were much, much worse years ago. They made the Cape and Islands, with their huge tracts of marshland, no place to be out of doors at night. Mosquitoes today, however, no longer are a hazard, only a mild summertime nuisance, as they are everywhere. For during the past two decades, the Cape Cod Mosquito Control Project has made large-scale and very successful efforts at control.

The line of attack is aimed at the marshes themselves, for these are the breeding grounds of the insects and the development region for the larvae. Ditches are dug across them and their creeks to drain them and make them too dry for the insects to breed. So far, about 2000 miles of such ditches have been completed in the marshes. Sod and plants are removed. The low areas in the marsh, which would hold tidal water, are filled in, and tidal gates are put up across channels so that larvae are starved of the daily flooding necessary to their survival.

When widely carried out, this is an expensive job, costing about seventy dollars for every acre in a three-year program. The cost is great, but the profit greater. Word spreads fast, and mosquito control projects on Cape Cod have been accompanied by a fifty percent rise in the tourist trade. The ditching of the marshes has removed one of nature's few public enemies from a land rich with her beauties.

CHAPTER XII

The Sea Draws the Map

The observer who from some lofty promontory surveys a broad sweep of interlocking bays and hills where rugged lands come down to meet the sea, finds in wave-cut cliff, sandy beach, and shingle bar abundant proof that he beholds an ever-changing scene. The shoreline stretching in intricate pattern on either hand is not the same shoreline as yesterday, nor the one which will border the lands tomorrow. The eye of the imagination, schooled in the principles of shoreline development, enables our observer to see not merely the fact of change, but better still the order of its progress. . . . In the most complicated pattern of the coastal margin may one trace successive phases of shoreline evolution, and so find in apparent chaos, order and the majestic beauty of natural law.

DOUGLAS JOHNSON
The New England-Acadian Shoreline

Now we turn from the quiet inlands of the Cape and Islands and look, for the remaining chapters, to the sea. For it is the all-pervading presence and power of the sea—thunder of breakers, sunny beaches, fog, gales, salty smell of tidal flats—that those who live on the Cape and Islands carry always in their minds. These narrow lands and all upon them today remain but incidents to the sea; and it is the sea which ultimately will take them.

It is the sea which really draws the map of Cape Cod and the Islands. Glaciers provided only the roughhewn lay of the land. If we compare maps of these lands drawn at fifty-year intervals we find that the sea goes at this work with the fierce dissatisfaction of a self-critical artist. Never has it made the finished drawing. Instead, it busies itself with a series of preliminary sketches, adding lines and curves here and there, extending the sketch in one direction, then another, changing the outlines and sometimes, in a seeming frenzy, erasing whole sections in one sudden sweep.

We needn't venture to guess at the shape of the finished picture. The artist has eternity with which to work. In fact, the sea may well completely erase the last of its sketches in this region, turn away, and begin a new creation elsewhere.

The Work of Shore Currents

Shore currents and waves have accounted for all of the land added to the original postglacial outlines of Cape Cod and the Islands. On a large-scale map of the Cape-Island area we can spot the directions of currents along the coasts and see the results of their work. Wherever there are sandbars, spits, necks and hooks, they trend parallel to the movements of the currents which built them. The building materials were picked up by the currents from shores which they passed; ultimately, therefore, from the morainal or preglacial substance of the Cape and Islands.

New Land on Cape Cod

Let us consider a map of the Cape elbow. Two long strips of land dangle from this corner. Nauset Beach reaches southward from Nauset Harbor to Chatham and creates Pleasant Bay. Monomoy grapples southward toward Great Point, Nantucket. Sands derived from the seaward edge of the Sandwich Moraine and the tablelands of Nauset go into the Nauset strip. As this grows, it too becomes subject to the attack of the sea and gives back some of its sand, which is carried still farther south to enlarge Monomoy.

The currents here are active. Nauset Harbor, which now opens into Town Cove, Orleans, was in Eastham three centuries ago. It is being pushed southward continually by a bar extending along it. Monomoy is an apt crazy bone for the Cape elbow. It has shifted and crept, divorced itself from the mainland and rejoined, and has grown as much as 175 feet in a single year, although its rate of growth is usually slower. For no two years does it remain the same. At one time, in the 1850s, Monomoy even produced a short-lived harbor, Powder Hole, large enough for forty ships. Today an enclosed body of water of that name, at the southern end of Monomoy, is all that remains.

It is unlikely that Cape Cod and Nantucket ever will be united in this area, strain as they may toward each other. When their respective spits reach out too far they hit a rapid east-west current which flows between them. This overpowers the shore current and swings its materials westward. That is why both Monomoy and Great Point are curved sharply, like hooks, to the west.

Down at the northern end of the Cape, the great Provincetown hook—"one of the best harbors on the Atlantic Coast," according to the *Coast Pilot*—is wholly the result of currents. Active tides and currents allow the waters around the harbor to maintain their twenty-fathom depth on its west side, even near shore; and they keep the harbor dredged to its nine-fathom depth between Long Point and Provincetown.

Southeast storms across the ocean have set up currents which have swept materials northeastward from the sea-eaten shores of Wellfleet and Truro. Meanwhile, on the Bay side, currents from the southwest have built up the narrow strip of land northeast of Pilgrim Heights, along which the old highway runs parallel to Pilgrim Beach. Since 1869, a man-made dike has joined these two spits farther west and enclosed the large lagoon, Pilgrim Lake (once East Harbor) between them (*Figure 41*).

Long Point owes its origin to currents from the northeast which, following the growing coast, have built it around to the east and southeast. Now, reinforced by a long, man-made rock dike, it protects Provincetown's "good harbor and pleasant bay, . . . wherein a thousand sail of ships may safely ride." So the Pilgrims described it, and since, during heavy storms, it has been said that the harbor actually has been host to a thousand ships. The rock dike serves to keep current-borne sand from filling in the harbor. Instead, the sand piles up on the west of dike and harbor. This has made the quiet water there shallow enough for the growing salt marsh which someday may fill it completely (*Figure 42*).

Provincetown's other, much smaller harbor is Hatches Harbor, just east of Race Point. Race Point is growing rapidly today, catching part of the coastal debris otherwise destined for Long Point. Less than a hundred years ago there was an open inlet, Race Run, behind Hatches Harbor. This since has silted up, and the dike across the head of the harbor breaks the influx of water from Cape

Figure 41. *At Cape Cod's sandy wrist. The magnificent dune country of Provincetown-Truro and Pilgrim Lake, a onetime harbor.*

NATIONAL PARK SERVICE

Cod Bay. So thoroughly has Race Run acquired a landfill that today it is the site of the Provincetown airport.

The entire Lower Cape from Orleans northward would be two islands today if the sea had not raised two narrow sandbars to close up schisms created by natural lowlands. Pamet River's valley is the northern of these breaks. It would be a natural canal across the Cape at Truro except for the current-built bar at its eastern end. This not only keeps out the sea but also allows for marsh growth that may fill Pamet's valley someday and permanently putty up the crack in the land. The other, southern lowland, as we have seen in Chapter VI, is Jeremiah's Gutter, or Boatmeadow Creek in Orleans. Seventeenth-century maps show the Cape as an island north of this creek (*Figure 29*). Although the sea still may break through in heavy storms, a sandy lowland now dams it in its eastern end, and marshes are filling in its channel.

A map of Wellfleet Harbor shows that the currents along the Bay coast here move to the south. They have piled up the sandbars which string together the "islands" in the harbor, adding a leftover string of sand hanging southward. The same currents have tied Indian Neck and Lieutenant Island, also in Wellfleet, to the land.

Barnstable Harbor, on the other hand, shows eastward-trending

Figure 42. *The long rock dike at Provincetown. This dike protects the harbor from sand which drifts toward it in the shore currents. The sand piles up on the right and provides a nucleus for marsh growth.* B. CHAMBERLAIN

currents, for Sandy Neck's dunes are built of the morainal deposits of Scorton to their west.

The excellent beaches on Cape Cod's southern shore along the Sounds appear as a line of white stitches of current-built sandbars and barrier beaches. They partly bridge the outwash channels near Falmouth, form Poponesset Beach near Hyannis, and connect Great Island to Yarmouth, forming Lewis Bay in the process.

On the Cape's Buzzards Bay shore, where the moraine swings close to the water, outlying glacial knobs became islands when rising seas drowned the shoreline. Currents since have woven narrow threads of sand from many of these islands, by which the lands drift from the Cape like balloons. Wing's, Scraggy, and Bennet's Necks, now attached to Bourne, were such islands once; as were Falmouth's Chappaquoit Point and Long Neck in Woods Hole.

New Land on Nantucket

Nantucket Island's Great Point, which reaches north from the island toward Cape Cod's Monomoy, has a glacial core, but its long hook is made of sand from the eastern headlands of Squam and Sankaty. Thus, the currents which build onto Great Point flow northward toward the entrance of Nantucket Sound. The presence of the Point itself has given rise to southward-flowing currents which are widening Coskata Beach and building up Coatue Beach. Coatue is a long bar, scalloped to perfection on the side facing Nantucket Harbor. Its graceful pattern is the work of tidal currents which alternate up and down the coast, piling sand first in one direction, then in the opposite (*Figure 43*).

Nantucket Sound breakers apparently once beat against the strip of high land which partly forms the eastern shore of Coskata Pond and the knobs of morainal material to its south. As the beach widened and the Great Point spit lengthened, Coskata Beach began to enclose and make a lagoon of Coskata Pond.

In shell heaps left by Nantucket Indians there is evidence that this entire region may have moved westward, including the head of Nantucket Harbor and Coskata Beach. The western edge of Coskata has been built since Indian days. On its southern shore, between Coskata and Coatue Point, there are layers of shells be-

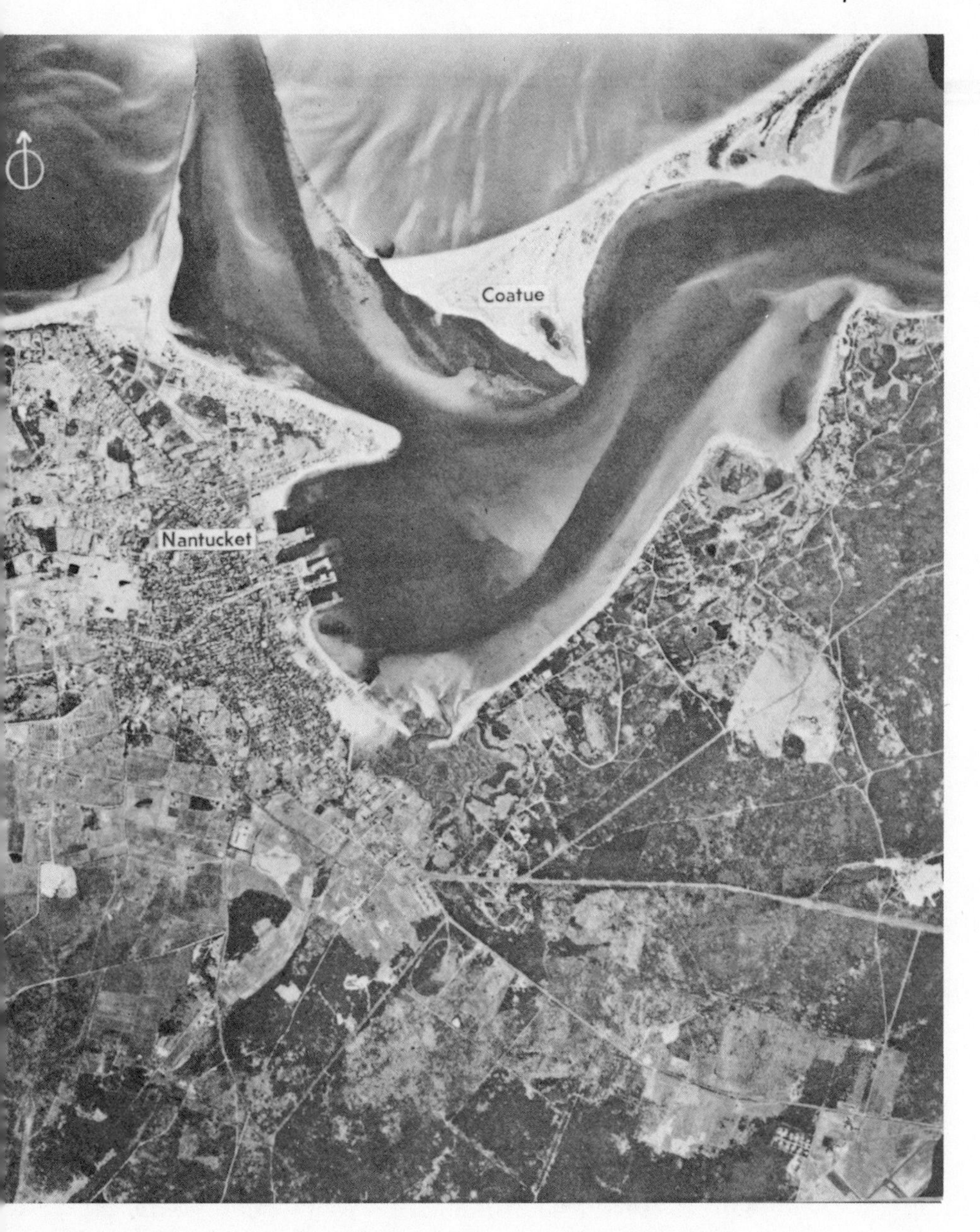

Figure 43. *Nantucket Town and Harbor, showing Coatue Beach. Coatue (upper left) is a wide, sandy spit scalloped by alternating currents. Long rock dikes keep the harbor open.* U. S. COAST AND GEODETIC SURVEY

neath the windblown sand. Doubtless the Indians placed their discarded shells right beside the large harbor as it was then. Excavations reveal older, lower layers of quahogs in these shell heaps, followed by later layers of oysters. At the time that the Indians were feasting on quahogs, the sea was eating into the south side of the Coskata glacial deposits. It obtained clay from these, and spread it over the harbor bottom, together with some pebbles sprinkled here and there. Such mud-covered shallows are the favored domain of quahogs, so that at that time, as the shells indicate, quahogs were the dominant shellfish.

Meanwhile, however, that part of the harbor now known as Head of the Harbor was creeping westward, for sand was drifting in from the ocean to the east. This sand finally covered the muddy harbor bottom. And thus the Indians' quahog supply stopped. They happily turned to gathering the oysters which were taking up residence in the harbor, attracted by the sandy bottom. On their shell heaps, oyster shells covered the clamshells, making the sequence that thus has shed light upon the history of the harbor.

Smith Point, which now is an island reaching westward from Nantucket Island toward Tuckernuck, has shifted and changed direction and location perhaps more than any other single area in the region. This restless strip of sand used to be the landing place for Indians traveling from Marthas Vineyard to Nantucket, and they must have had to revise their sea route frequently. Sometimes Smith Point has run all the way along the southern coast of Tuckernuck, parallel to it and separated from it by a long, narrow body of protected water. Other times it has gone still farther to extend west of Tuckernuck. Sometimes Smith Point has merged with Tuckernuck itself, only to have the tie broken again by heavy storms.

New Land on Marthas Vineyard

Southern Marthas Vineyard is a classic example of the abilities of waves to mend a torn and ragged shoreline. Its entire southern coast, the frayed edge of the outwash plain, has been neatly hemmed with a long barrier beach, South Beach (*Figure 13*). This ends in the eastern spit of land, Norton's Point, which partly encloses Katama Bay. Since the currents here flow eastward, some of these sands of South Beach have come from the outwash plain itself, where it borders the beach.

South Beach is a product of the waves. Debris from its bluffs, as from all soft sediment attacked by the sea, has found its way seaward to form a shelf of sediment which has widened continually with the wasting of the land. In the early stages of erosion, that shelf slowed up storm waves which charged landward. Thwarted, they scoured into the shelf itself, deepening it where their action was strongest into a sort of long basin. They threw landward the by-products of this scooping, forming a bar with the coarser sands. On the seaward side of this bar, waves continued to plaster material as they surged in. This caused the bar itself to sidle seaward. Finally, however, the offshore basin became so deep that heavy waves could sweep freely across it and concentrate their attack on the bar which earlier waves had built. The newly severe erosion removed material from the seaward side of the bar and carried it to the lee. This forced the bar to about-face on its journey over the sea floor and to start moving landward.

Thus, in time the entire strip was shoved backward until it stood at bay against the land. This final step may have occurred in historic times. At least there is a Vineyard tradition that in the eighteenth century a young man could put on his ice skates at Tisbury Pond and skate all the way to his ladylove in Edgartown on the long, narrow bay which then lay between the offshore bar and the outwash plain.

Along Chappaquiddick's east shore, another barrier beach ties the island to its northern tip, Cape Poge, which is itself a barrier beach. Thus the Poucha Pond-Cape Poge Bay body of water really is a lagoon, formed by these sandy, dune-covered, protective strips. As Cape Poge grew from sea-carried debris, it too became a source of sediments which now are building up the southward-stretching hook, Cape Poge Elbow. Someday the elbow probably will touch Cape Poge Gut, the beach along Chappaquiddick's North Neck, and will close off Cape Poge and make it a pond. This has happened already to Sengekontacket Pond, north of Edgartown, although the pond is kept open artificially. It is happening to Lagoon Pond in Vineyard Haven, too, for Lagoon Pond once was part of Vineyard Haven Harbor.

Over on the western coast, Gay Head's cliffs crumble to supply sand which, as it has closed off Squibnocket Pond in the past forty years, will one day close off Menemsha and Chilmark Ponds, unless they continue to be kept open artificially. Chilmark Pond's natural dam must be breached several times a year to keep the water salt. Menemsha has large jetties extending from it to deflect sand-bearing currents from filling it in.

Birth and Death of Harbors

The Cape and Islands abound with examples of how wave and current sand deposits can create harbors large and small. As we have seen, such major harbors as Nantucket, Barnstable, Wellfleet and Provincetown are protected by sand deposits. But there can be too much of a good thing, and the sandy barriers may so grow as to render the harbor useless again.

This happened on Nantucket. Copaum Pond, on the island's north shore west of Nantucket Harbor, was a good harbor itself once, sheltered from but open to the sea by a discontinuous bar. During the seventeenth century a large number of Nantucketers lived here, relying on the harbor for their maritime activities. The bar had been built by currents; and through the years, townspeople watched powerless as similar currents filled their harbor entrance with sand. In 1720 it became fenced off completely from the sea, and so it remains today. Their town thus useless, the inhabitants packed up and moved east to settle around the larger harbor of today's Nantucket Town. A memorial of the days when Copaum's shores were bathed by the sea is the tiny spit of sand jutting out from its eastern shore, a sea spit with its development halted by separation from the sea.

Today, of course, modern engineering is an opponent worthy of the powerful sea and its caprices. The threat to Provincetown Harbor from sand filling on its west has been controlled by the long rock dike between the town and Wood End (*Figure 42*).

However, the attack on the east continues as it did in 1893, when Shaler wrote: "It is a regrettable fact . . . that this haven is continually menaced with destruction by the assaults of the sea on the thin strip of sand which separates it at one point from the open ocean on the east. It is not impossible that one great storm . . . may break through the frail defenses at this point and fill up the harbor with a vast amount of sand, which would be driven into it as a necessary consequence of such a catastrophe."

Currents and Climate

The sea alone cannot take full credit for shaping the younger parts of the Cape and Islands. They are really the products of a working partnership of sea and wind, a business where lands rise and fall and fine beaches are but a luxurious sideline. Both members of the partnership are necessary. The role of the sea is obvious, but let us consider the importance of wind to its action.

Without wind, the only destructive capabilities of the quiet water would be painfully slow dissolving of mineral substances where water laps the land. The Cape and Islands would be larger, for wind-driven seas have eaten away more land than they have replaced. Without wind, the Cape and Islands would have no beaches, no sea cliffs, no terraces, bars, lagoons or spits. The only constructive activity of the sea would be the passive reception of sediments on it shallow floor from streams of the land.

The climate and even the life of the Cape and Islands depend somewhat on the sea-wind team. Cape Cod, a peninsular Janus, faces two completely opposite ocean currents. Its northern half, that part along Cape Cod Bay and the narrow arm of the Lower Cape, faces the cold Labrador Current which flows down the continent's eastern coast until it nearly reaches the Cape. The other side of the Cape—the Vineyard and Nantucket Sounds side—faces the influence of the warm Gulf Stream, which comes up the coast from the south and swings eastward just south of Cape Cod.

These currents, originating in a chain of wind, water temperature and earth rotation reactions, make Cape Cod Bay's temperature about ten degrees cooler than the temperature of the Sounds. Thus, the Cape is a faunal boundary. For years, its curved arm has embraced within its crook a sea population of northern species, while just a few miles away on the opposite side, southern forms have dwelt. This has provided an exceedingly rich assortment of life—more than a thousand species.

The animals peculiar to each side are wholly unwilling to live in the domain of the other. "The difference [in flora and fauna] between Cape Cod Bay and Buzzards Bay is greater than that between Cape Cod Bay and the Bay of Fundy or between Buzzards Bay and the coast of Virginia," states a National Park Service report. Along the Cape's ocean and Sound sides we find such warm-water seashells as certain conches, transparent jingle shells and fluted ark shells. In Cape Cod Bay are cool-water snails and periwinkles. Other creatures are more adaptable and can be found in both regions. In the last century, Gould reported that of 197 species of sea life studied here, eighty-three were found only on the north side of the Cape, fifty only on the south, and sixty-four on both.

Since the 1930s, however, worldwide climate has moderated. Southern species have been nudged northward, so that today there is more overlap. Thus, Cape Cod and the Islands are fishermen's delights; for in their open waters mingle warm-water mackerels, bluefish, tuna, swordfish, sharks and striped bass, with cold-water cod, flounders, eels and halibut. The northern types are drifting away, however. Cod, which once swam "in great plentie" around these shores, have forsaken their namesake land for Greenland's chillier waters.

A temperature gradient between equator and poles creates the Gulf Stream. Cold water flows beneath the surface to the equator, replacing warm water, which heads poleward on the surface. The course of the Gulf Stream is charted by the earth's rotation and the trade winds which throw it against South America, from which it heads north. This river of warm water is up to sixty miles wide in places, more than 6000 feet deep, and flows up to two and a half miles an hour; we could walk unhurriedly alongside it on the land and keep up with it. From just south of Cape Cod the Gulf Stream crosses the Atlantic to take warmth to Great Britain and Scandinavia.

It is perhaps common knowledge that among Benjamin Franklin's many accomplishments was our first chart of the Gulf Stream, drawn up in the mid-eighteenth century. Less widely known is the fact that it was Nantucket whaling men who gave Franklin much of his information and a rough draft of his chart. These whaling men pursued their livings along the very edges of the ocean current as far south as the Bahamas, and probably knew as much of its speed and size as anyone then living. As Franklin said in a letter dated 1769: "At my request Captain Folger [of Nantucket] hath been so obliging as to mark for me on a chart the dimensions, course and swiftness of the stream from its first coming out of the gulph [of Mexico] where it is narrowest and strongest, until it turns away to the southward of the western islands, where it is broader and weaker, and to give me withal some written directions whereby ships bound from the Banks of Newfoundland to New York may avoid the said stream; and yet be free of danger from the banks and shoals [Cape Sable Shoals, Georges Bank, and the Nantucket Shoals]."

Thus, among the products of the sea-wind partnership are ocean currents, where the unchanging trade winds react with the sea in its temperature circulation to carry climate. It is the lesser movements of the sea, however, induced by the gypsy winds of local weather, that carry sediments and create and destroy lands.

Wind and Waves

Hector Crevecoeur stood entranced on the shore of Nantucket in the early 1780s, facing the sea-borne wind:

> . . . and who is the landsman that can behold without affright so singular an element, which by its impetuosity seems to be the destroyer of this poor planet, yet at particular times accumulates the scattered fragments and produces islands and continents fit for men to dwell on! Who can observe the regular vicissitudes of its waters without astonishment; now swelling themselves in order to penetrate through every river and opening, and thereby facilitate navigation; at other times retiring from its shores, to permit man to collect . . . shellfish . . . ? Who can see the storms of wind, blowing sometimes with an impetuosity sufficiently strong even to move the earth, without feeling himself affected beyond the sphere of common ideas? Can this wind which but a few days ago refreshed our American fields, and cooled us in the shade, be the same element which now and then so powerfully convulses the waters of the sea, dismasts vessels, causes so many shipwrecks, and such extensive desolation?

Miles east of Sankaty on the open Atlantic, at this moment even as in Crevecoeur's time, air is moving against the surface of the water, dragging it up by friction. The wind is generating a wave. The size of the wave depends on three things: how long the wind blows; the speed of the wind; and the distance it has traveled unhindered over open sea. Once the wave is born, the pulse of energy, embodied in the wind that gave it birth, pulls it along by friction and pushes against its sloping side. The wave now moves over the surface of the sea.

The wave moves, but a piece of wooden drift lying on the water remains nearly in place. Perhaps it inches forward just a bit, but mainly it rises and falls to let the wave pass beneath it. The water supporting the wood thus moves differently than the wave. Let us trace the actual movement of the water itself.

Laboratory studies show that as a wave moves horizontally across the water surface, the water travels in vertical circles. When we watch waves generated artificially in a tank in which there are bits of seaweed, this is what we should see: the seaweed, caught up in the water over which the wave passes, makes vertical loops in the water. At the crest of the wave, the portion of water carrying the seaweed is at the top of a loop, traveling in the direction the wave is moving. At the trough of a wave, the same portion of water with its seaweed is at the bottom of a loop, moving opposite to the direction of wave movement. A wave also affects the water beneath it. The water highest in the tank transmits its vertical looping movement to the water below, which moves in smaller circles and so on down, until at depth, the water moves in tiny circles, then ceases altogether.

The surface water, given a direct impetus by the wind, moves a little faster than that beneath; the lower water's movements are weaker echoes of the higher. Thus, the surface water itself may be carried forward slightly, because of wind drag, with the wave.

Now we can leave the experimental tank and go back to our hypothetical wave east of Sankaty, where the same sort of thing is happening. The wave has gathered itself together in the ocean and now is traveling toward the coast, surrounded by a train of other waves. Let us follow in this wave train, or swell. Any individual wave will not survive a long journey, for it gradually weakens and dies; but other waves are born to take its place in the swell.

Wherever the waves pass over the water, their pulses set water particles rotating, and this happens nearly all the way into shore. When the waves reach shallow water, the vertical loops of moving water reach down all the way to the sea floor.

By now the wave has reached the gently sloping coastal shallows. As we watch successive waves rolling in, we see an interesting phenomenon; they bend and adjust themselves, near shore, to the pattern of the coast. Perhaps they started their approach at an angle rather than with their fronts parallel to shore. But those parts of the wave front nearer shore feel bottom sooner and slow up by friction, while the segments farther out continue full speed ahead until they catch up. In this way, a wave which started toward shore at an angle tends to realign itself more parallel to the coastline.

The wave is right before us now. We can watch it build up. It reaches the place where its height is somewhat less than the depth of still water. This throws the water into confusion, because no longer can the water loops travel their complete circles. Quickly then the wave appears to get higher; it steepens further; the crest heaves itself up; moving ahead of the rest of the friction-slowed water, it becomes top-heavy; and it crashes in foam and spray, into its own trough. The wave has ended its long trip by becoming one of the endless breakers which pound up and down the beaches at Nantucket.

If we wade out into the surf at the beach by Sankaty where our wave has just broken, we might see a sharply defined ditch along the breaker line. This is the product of constant hammering by successive lines of such breakers. Even if we fail to see the ditch, our feet will find it, for it is sure to be a catchall for coarser pebbles drawn seaward with the backwash.

If we watch the line of breakers for a while, they will tell us some facts about themselves. Several rows of breakers with foaming whitecaps beyond indicate that they are produced by local onshore winds. But waves which strike shore as a well-defined line of sharp, steep breakers probably started their journeys far away, originating in some far-flung storm.

Scanning the breaker line from clifftop, we see gaps, one or two hundred feet wide, which interrupt it, places where the surf is not as high or as active as elsewhere. A bather might think such a break a good, quiet place to swim, but let him approach with extreme caution. Such breaks are the passages of rips, or riptides—an unfortunate term which has nothing to do with tides. These rips are the channels where piles of wave-driven water converge to find their way back to sea, sometimes at a rate as fast as more than a mile an hour. A swimmer easily can be carried out to deep water here, and a great many drownings blamed on the mythical undertow actually are caused by rips. Swimming against a rip won't work. The only way out of one is to swim parallel to shore and thus get beyond the rip.

The Raising of Lands

Let us turn our attention to the seaweed, shells and stones at the waterline—all the material carried or rolled by waves as they strike the land. With these we may trace the path of the water after a wave breaks. We notice that some of the suspended objects slide directly back down the beach as backwash; others hurry out to sea via rips; and others move parallel to shore just beyond the breaker line. These are in the grip of a longshore current. Such currents are the prime builders of Cape-Island postglacial additions.

If we suppose that the sea wind blowing upon a north-south coast is from the northeast, then those drifting scraps are being carried southward. For the waves meet the coast with a slight northwesterly-southeasterly slant, and each wave, when it breaks, splits into two components of movement. One is the backwash, which goes directly back to sea. The other component is what moves the flotsam southward in the water, for this component runs southerly, parallel to shore, and is the longshore current. The net result—set up by thousands of similar breakers—is a narrow current of water moving along the coast.

Longshore currents flow before the wind and parallel to the shore. If we try swimming beyond the breakers under such conditions, we find we make much better progress going south than north. For, as many surf swimmers know, such a current is an actual stream of water, unlike a wave's simple passage of energy across the water's surface. It may flow a mile per hour, especially if reinforced by a wind parallel to the shore.

It takes a series of waves parallel to one another to set in motion a longshore current. Thus, such currents are common only where there are prevailing winds from one direction.

If, however, disturbed weather conditions cause waves to approach shore from varying directions, they run together, piling water into a temporary peak. The peak must collapse, and in so doing can send a current moving in any direction at all. If we go back to Sankaty's beach at low tide we may see the work of such unorganized currents. They shove offshore debris farther out, leaving a scoured zone toward shore and helping to build a bar beyond.

On most parts of the Cape and Islands, longshore currents are active and noticeable. While Henry Beston lived for a year on the dunes of Eastham, Cape Cod, in his Outermost House, south of the Coast Guard station, he took note of the direction of the currents in the sea: "Shore currents here move in a southerly direction; old wreckage and driftwood is forever being carried down here from the north. Coast Guard friends often look at a box or stick I have retrieved and say, 'Saw that two weeks ago up by the [Nauset] light.'" Had the wood continued on its way south, Orleans Coast Guard men eventually might have noticed it drifting by as had the men earlier at Nauset Light. Still farther it might have gone, tied to the current another ten miles, perhaps, finally to be snagged by Monomoy. Because of the general southward trend of the currents here, a great deal more debris gets taken to Monomoy than north to Provincetown from the sand plains.

No one who has walked much along beaches can fail to have noticed the festoon pattern of the wrack upon the sand. Many of the sand grains near the edge of the beach are arranged likewise, but their pattern is well disguised among all the other sand grains. This scalloped trail of ocean castoffs is the combined work of incoming waves, backwash, and longshore currents.

A wave which meets the shore at an angle, however slight, pushes up sand and gravel, flotsam and jetsam along its front. The breaker's longshore component carries along the coast any debris left in the water. Meanwhile, the water which returns to sea as backwash goes straight back, perpendicular to shore, and drags with it some of the sand and wrack. Locally, therefore, materials move in a scalloped path, zigzagging along the beach, but ultimately they make headway parallel to it.

Much of the sand on which we stood to watch the breaking waves at Sankaty had, like Beston's driftwood, already traveled miles and would travel miles farther. It would stop, perhaps, at some sandspit trailing off Nantucket—or perhaps even float somewhat out to sea.

The reserves of power behind this simple process are truly amazing. Huge rocks amble down the beaches, parallel to the currents. Years ago, the heavy anchor and chain of a large ship lay on the edge of the Truro Outer Beach, Cape Cod. In three weeks it had wandered a mile and a half northward from its original position on the beach.

Sometimes, however, rocks move along the beach contrary to

the general movement of sand, wrack and currents, breasting the crowd on a one-way highway. By this "singular paradox," wrote Woodworth, "on certain beaches, the southwesterly dry wind moves the sand northward and seaward, so that when the wind is strong the beach is lowered several inches during a single low tide; but at the same time the pebbles on the beach work up against the wind in the opposite direction to that in which the sand is traveling, for the wind scours out a small pit on the windward side of each pebble, into which it slides under the action of gravity as the sand is blown away from around it. In this manner pebbles work their way from the headlands along the wasting beach independently of the longshore motion of the surf."

The great work of these currents parallel to shore, as we have seen, is to build up lands by carrying sediments along the coast. Lands spring up most commonly where the direction of the shoreline changes abruptly. In such a place a current cannot change swiftly. Its inertia keeps it in a straight line. So beyond the land into deeper water it goes, and there it drops its burden. This forms a spit, trailing from the land in the direction of current movement like a kite's tail in the wind. Such is Cape Cod's Monomoy. If the spit runs across an open bay it becomes a bar or barrier; bars putty up the dents of a coastline, like Marthas Vineyard's South Beach. Along coasts where headlands and bays alternate, local currents may form which sweep sands to the bay mouths in curved spits; these are bayhead bars. One of Cape Cod's bayhead bars, Craigville Beach, Hyannis, has been classed as one of the four finest beaches in the world.

Jetties and Their Relatives

Longshore currents frequently are put to use to remedy the ills of bathing beaches. Any obstruction that slows up a current causes it to deposit some of the sand which it carries. Thus, rock groins and jetties have been set up of late along populated parts of the bays and sounds of the Cape and Islands.

Customarily, "jetty" refers to a wall of rock built at the mouth of a creek, "groin" to a wall of rock which extends out from a shore beach. These current baffles reach out perpendicular to the shore from near high-tide mark and waylay sand which moves down the coast with the current. Soon crescents of sand form in the current-facing corner between the groin and the shore, and spread out to widen the beach. On Cape Cod's southern shore from Falmouth Heights eastward there are some excellent beaches enlarged by groins and jetties.

Whoever builds the first groin on a beach is starting a chain reaction which may lose friends for him and may ultimately lead to a rock-ribbed beach to leeward. For you cannot use a groin to build up Beach A without simultaneously robbing Beach B downcurrent from A. Beach B, its sands washed away and no longer replaced, soon starves into a coarse tract of left-behind gravel. At this point the owner of Beach B waxes wroth, builds his own groin if he can afford it, and thus loses the friendship of the owner of Beach C, on his downcurrent side.

Jetties are protective devices for harbors as well as for beaches. Many are of prime importance to the passage of ships in and out of harbors by keeping their entrances clear. Two jetties protect the entrance to Bass River, but in recent years the approach to the jetties themselves has shoaled, and a channel has had to be opened between them. The entrance to Witchmere Harbor is kept open by a long, hooked jetty.

Dikes are similar to groins and jetties in principle, but larger. Usually they are built to protect coastlines or to keep harbors from sanding up. A good example of this is at Provincetown. There were once two harbors for this town, the present Provincetown Harbor and East Harbor, which is now Pilgrim Lake. During the nineteenth century, as waves threatened to cut through the narrow wrist of Truro, a dike went up to protect it. East Harbor was the sacrifice; the dike severed it from the Bay and turned it into a lagoon (*Figure 41*).

Two long rock dikes keep open Nantucket Harbor (*Figure 43*). Before they were built an underwater sandbar stretched across the harbor. Known to islanders simply as "The Bar," it was long a hazard to ships entering and leaving the harbor. In the late eighteenth century a dry-dock contraption known as the "Camels" was used to float ships over The Bar. *Old Ironsides* was its first customer.

Breakwaters are rock walls built parallel, rather than perpendicular, to a coast, to protect an open harbor from storms. Cape Cod's Hyannis Harbor is partly maintained by one. Brewster, on Cape Cod Bay, has no natural harbor at all. To provide one for the Boston packet ships during the eighteenth century, the town built a long rock breakwater which bore the brunt of storm waves. The barnacle-encrusted remains of the breakwater still emerge at low tide, a short distance northwest of the Brewster town landing.

Tides and Tidal Currents

Finally, there is one special sort of water movement produced along the coast. No one who visits Cape Cod or the Islands can fail to notice it. This is the movement of the tides. Tides on the Cape Cod Bay shore average about ten feet; those at Nauset Harbor, six and a half feet; those on the south shore, two feet; and those on the Islands, a little over one foot. These tidal movements can behave like any waves and currents, but they come and go with the regular pulsation of the tides.

Twice a month, of course, tides reach their highest levels; once when the sun and moon line up to pool their gravitational pulls on the earth's waters, and once when the earth lies directly between moon and sun. These are the spring tides (no relation to spring as a season), the tides of the new and the full moon. Neap tides are the two smallest tides of the month. They occur when sun and moon form a right angle with the earth during the moon's first and third quarters and offset each other's pulls.

Actually, tides are not simple diurnal affairs. They vary considerably with the topography of the sea floor, with the shape of the surrounding lands, and with the seasons; so that even on opposite sides of the Cape we find tides differing by as much as eight feet.

Waves and currents of the tides, like wind-produced waves and currents, can carry sediment and even move pebbles as deep as 800 feet below the water's surface.

In 1869 the director of the United States Coast and Geodetic Survey reported, concerning the amazingly strong tidal currents around the Islands: "There is no other part of the world, perhaps, where tides of such very small rise and fall are accompanied by such strong currents running far out to sea." The Vikings, in the eleventh century, gave the name *Straumey*, referring to an island with strong currents, to what may have been either Nantucket, Marthas Vineyard, or Monomoy.

Such tidal currents can be exceedingly swift and deep where they plow through narrow gateways of the land. Oceanographer Francis Shepard has described his frightening experiences with such a current off Woods Hole some years ago:

> In the old days before the channel was adequately buoyed, one could have a harrowing experience going through Woods Hole near the Cape Cod Canal. I remember one such trip on a sailboat during a spring tide. We had to anchor at the western entrance of the Hole until the tide turned in our favor. By the time we got sails up and were underway, the favorable current had picked up such a velocity that the buoys were actually carried under water and could only be recognized by the swirling eddies around them, which did not differentiate them from the rocks. The channel was so crooked and the speed so great that it was almost impossible to keep our position on the chart. All we could do was to avoid all the larger eddies and hope that this would keep us from colliding with the rocks. We dodged back and forth, and after a few hectic moments we were through the narrows and sailing rapidly out into Vineyard Sound at the far end. The placid waters of the sound were a welcome sight.

The scarcely tapped power of the tides has been used on the Cape and Islands to power mills. Dams caught the high-tide waters, later releasing them to drive mill wheels. One such mill ran for a good many years at Pocasset; another was operating at Wellfleet in 1790. Despite all the power of the tides, however, their sediment-rearranging effects along the shore zone go scarcely noticed amid the more lively activities of the unchainable wind-driven waves.

CHAPTER XIII

Dangerous Shoals and Roaring Breakers

. . . . they tacked about and resolved to stand for the southward to find some place about Hudson's River for their habitation. But after they had sailed that course about half the day, they fell amongst dangerous shoals and roaring breakers, and they were so far entangled therewith as they conceived themselves in great danger; and the wind shrinking upon them withal, they resolved to bear up again for the Cape [Provincetown]. . . . That point which first showed those dangerous shoals unto them, they called Point Care, and Tucker's Terror; but the French and Dutch to this day call it Mallebarre, by reason of those perilous shoals and the losses they have suffered there.

Governor of Plymouth, WILLIAM BRADFORD
1630–1650

As the sea attacks old lands and raises new, it must build by slow degrees, up from the sea floor. Wherever this land-raising work by wind and sea is active and continuous, as it is around Cape Cod and the Islands, the offshore waters become a challenge to the skill of sailors—and sometimes a stage for tragedy.

Types of Shoal

The great treachery of such waters lies in the vast miles of hidden shoals, new lands forming which have not yet broken through the sea's surface to the open air. Provincetown's Long Point, Chatham's Monomoy, the Vineyard's Cape Poge, Nantucket's Great Point—and all similar stretches of water-carried sand—are onetime shoals which, having lurked undefined and dangerous beneath the sea's surface, now lie unmasked in the light of the sun.

The offshore waters of the Cape-Island region are full of such nearshore embryo lands. In addition, the area's submarine contours are complicated by another type of shoal, old lands which have died, drowned in earlier earth ages. These include onetime islands and the submerged glacial moraines of Georges Bank and the Nantucket Shoals.

And finally, superimposed on these relatively mappable parts of the seascape, the sea hourly provides changes of greater immediacy to harass those who sail. Charts of coastal waters scarcely can keep up with them. The Coast and Geodetic Survey in its surveys of the region must concentrate its efforts on those areas important to commercial shipping and those having depths critical to all navigation. In the rest of the waters, the Coast Guard is the agency which must continually watch the changing shoals and move its buoys accordingly. These moves necessitate frequent revision of the Coast and Geodetic Survey charts. Revisions of the chart of Nantucket Sound and its approaches have been published fifteen times in the past ten years.

Endlessly changing windblown waves, manes flying, pounce like lions on the strand materials, tossing, teasing, rolling them from one place to another; chasing, catching, teasing again. Bars sidle and shift, become islands and fall back; shoals break and form; sea cliffs crumble and sand necks split apart. Timbers of wrecked ships vanish and reappear like ghosts. Within a year the shifting sands may hide from view the wreckage of a ship battered to destruction on those very sands.

Shoals in Cape Cod Bay

In Cape Cod Bay the worst shoal area is off Wellfleet. The currents here which have enclosed Wellfleet Harbor with a series of sandbars also create shoals. These are further complicated by sea-swallowed lands such as Billingsgate Island. Because of the string of hidden hills southwest of the harbor, boats from the north sometimes must go south and make a hairpin turn to enter the harbor safely.

Shoals in Nantucket Sound

Nantucket Sound is twenty-three miles east to west, and from six to twenty-two miles wide. In most of this substantial body of water, shoals are the rule and deep waters the exception. Many and varied are the hidden lands of the shallows—growing spits, drowned islands, shifting bars. Inexperienced sailors are well cautioned away, far away, for there are only a few charted channels through the intricate maze. The second lighthouse of the United States went up on Nantucket in 1746 to warn ships of the submarine bars north of the island.[17]

In addition to the current bars, this entire region is one of hidden ice deposits, and this, of course, means rocks. Sometimes huge boulders are perched upon the near-surface hills. Along the shores of both sounds are countless glacial boulders, indications of what lies hidden farther out. For instance, a boulder-studded shoal, Devils Bridge, with a buoy marking its end, stretches westward at least a mile from Gay Head. One hundred twenty-one persons lost their lives here on January 18, 1884, with the foundering of the ship, *City of Columbus.*

Large rocks, some unmarked, lie hidden in Hyannis Harbor. Half a mile southwest of the end of the breakwater, for example, is the Eddie Was Rock, covered by only five feet of water and unmarked. There are three boulders strung across the shallows from Point Gammon to Hyannis Harbor: Great Rock, marked by a beacon; Gardiner Rock, marked by a buoy; and Halftide Rock, unmarked.

Finally, here as in all shoal-filled water, the currents which create many of the danger areas add to the hazard by their speed and, in some cases, the unpredictability of their moving waters. At Nantucket Sound's northeastern end, where it meets the Atlantic Ocean off Chatham and Monomoy, there is added treachery in the currents which are rapidly extending Monomoy and Nauset Beach. The only passage through the wide belt of shoals which corrugates the sea floor between Monomoy and Great Point runs, shiftingly, through the center, between Point Rip (the northeastern underwater continuation of Nantucket's Great Point) and Great Round Shoal to its north. This probably is the most actively changing section of the entire coast. At low tide, only small boats can get through from Chatham to Pleasant Bay, for the entire passage is shoal.

[17] The first lighthouse was that of Boston, built in 1715.

Shoals between the Islands

Novices had better not try the inter-Island route either. West of Nantucket is extensive Tuckernuck Bank, from which Tuckernuck and Muskeget Islands rise. Sometimes, in fact, wave-raised beaches connect Tuckernuck to Nantucket Island. Farther, beyond the bank, is a series of shifting underwater bars. Wasque Shoal, south of Chappaquiddick, has some areas which emerge from the sea at low-water stages. Currents, particularly tidal currents, are treacherous and active around the Islands. In Vineyard Sound between Marthas Vineyard and the Elizabeth Islands, currents have thrown up an extensive shoal known as the Middle Ground. Travelers passing on the ferry between the Islands can see the Cross Rip Lightship marking the Cross Rip Shoal, which rises to only eleven feet beneath the water surface.

Shoals off Chatham

Where the Sound meets the ocean near Chatham, the sea indulges in some of its liveliest frolics along the entire Atlantic coast. Debris piles into this region, driven up the Cape's eastern coast by the winds of northeast winter storms. The area east of Monomoy is shoal in some places as much as a mile and a half out. These hidden, ever-changing bars one day will rise above the surface as extensions of the land's long southward-hanging spits. Even Monomoy itself is sometimes attached to Chatham, sometimes an island under the influence of these unsettled currents.

The sea occasionally breaks over Nauset Beach. After an 1871 gale, Chatham Light faced open water. Vessels are warned by the *Coast Pilot* not to approach the shore in this region nearer than one mile for four miles south of Chatham Light.

There are three lightships off Monomoy, each with a red hull, to mark the worst danger areas and the passages through them. Seven miles offshore, Pollock Rip Lightship indicates the entrance to the Pollock Rip channel through the shallower areas. Handkerchief Lightship floats southwest of Monomoy Point to mark Handkerchief Shoal, which looms up from the seascape sudden and cliff-like on its southeastern side and is growing constantly. About a mile off Monomoy, Stone Horse Lightship marks Stone Horse Shoal.

Incidentally, in reference to the "eminently curious name" of this last shoal, Shaler wrote in 1897:

I am told by Captain John L. Veeder, of Woods Hole, that some years ago he was engaged in breaking up the wreck of a ship which had been for some time lying on that shoal and had become embedded in the sands. When the hulk rolled over it brought up a quantity of "dark sand" which contained many fragments of bones. In answer to my inquiry Captain Veeder stated that the material was like the greensands of Gay Head. It is well known that sailors are apt to class any bones as those of horses.

So active are the sand movements in this region that the federal government has had to initiate channeling projects to keep the harbor and passages open. One important proposed project (not yet in effect at this writing) is to open up the waters between Chatham Roads, which lie between the shoals of Monomoy and those along the southeast shoreline, and Stage Harbor, Chatham, the eastern end of which is slowly choking with sand.

Mooncussers

In the days of sailing ships, word got around among sailors that there was man-made treachery afoot on this coast, over and above the treachery of nature. There were tales of professional wreckers, especially around Chatham's feared waters. It was said that they created false lighthouse beacons by lighting fires on moonless nights, to lure ships onto the shoals so they could reap a harvest of cargoes. Cursing bright moonlight that revealed their activities, these men became known as mooncussers. But never was there an authenticated case of such villainy actually occurring on Cape Cod. To be sure, people did profit from shipwrecks simply by plundering stranded vessels, which, after all, at that stage were merely large-scale pieces of tidal wrack. A keeper of the first Nauset Light, in fact, reported to Ralph Waldo Emerson that certain local residents had strongly resented the presence of a beacon which would harm the wrecking business.

Such residents, and the mooncussing legend, have come down in history as faint smudges on a shining record. For Cape Codders traditionally have played a compassionate and heroic part in the rescue and sheltering of those whom the sea has tossed upon their shores.

The Ocean Bars

For years, Cape Cod's oceanside from Chatham to Provincetown has been known as the Graveyard of Ships (*Figure 44*). "The annals of this voracious beach!" said Thoreau; "Who could write them, unless it were a shipwrecked sailor? How many have seen it only in the midst of danger and distress, the last strip of earth which their mortal eyes beheld. Think of the amount of suffering which a single strand has witnessed! The ancients would have rep-

Figure 44. *Victim of the ocean bars. The hidden bars off Cape Cod's Outer Beach are infamous for their record of shipwrecks.* CAPE COD PHOTOS

resented it as a sea monster with open jaws, more terrible than Scylla and Charybdis."

The timbers of more than 3000 vessels lie buried in the offshore sands on the Cape, and most of them along this outer coast. The shores of Chatham alone—a few miles of sandy beach—are said to have received half the wrecks of the whole Atlantic and Gulf coastline of the United States. Peaked Hill Bars of Provincetown, offshore halfway between Race Point and Cape Cod Light, have been the cause of more wrecks than have occurred at any other part of the Atlantic seaboard.

The great killer all along this coast is the double line of such bars, hidden and restless, which the sea has thrown up parallel to the shore from Monomoy to Race Point. They are not always present in the sparkling days of summer. But in the winter, when winds are stronger and storms fiercer, the bars make matters worse by growing large and formidable.

They result from the rough action of winter waves, which load themselves with beach sands, pull back, and drop them in ridges behind the breaker line, providing peril for those at sea. The gentler waves of summer shove sand shoreward, planing the bars flat once more and building the beach seaward to provide pleasure for those on shore.

When an onshore wind blows, the surf foams over these bars. Often the presence of a bar causes waves to crash down on it before continuing, partly spent, to the actual shore. We may even see the breakers out there from the bluffs. As sailors have been well aware for centuries, there is little hope for a sailing ship storm-driven toward this coast. In 1802 Reverend James Freeman of Cape Cod wrote a description of these shoals for the benefit of the many mariners who had to pass the Cape coast.

Along the shore, at the distance of half a mile, is a bar; which is called the *Outer* Bar, because there are smaller bars within it, perpetually varying. This outer bar is separated into many parts by guzzles, or small channels. It extends to Chatham; and as it proceeds southward, gradually approaches the shore and grows more shallow. Its general depth at high water is two fathoms and three fathoms over the guzzles; and its least distance from the shore is about a furlong. Off the mouth of Chatham Harbor there are bars which reach three quarters of a mile; and off the entrance of Nauset Harbor the bars extend a half of a mile. Large, heavy ships strike on the outer bar, even at high water; and their fragments only reach the shore. But smaller vessels pass over it at full sea; and when they touch at low water, they beat over it, as the tide rises, and soon come to the land. If a vessel be cast away at low water, it ought to be left with as much expedition as possible; because the fury of the waves is then checked, in some measure, by the bar; and because the vessel is generally broken to pieces with the rising flood. But seamen, shipwrecked at full sea, ought to remain on board till near low water; for the vessel does not then break to pieces; and by attempting to reach the land before the tide ebbs away, they are in great danger of being drowned. On this subject there is one opinion only among judicious mariners. It may be necessary however to remind them of a truth, of which they have full conviction, but which, amidst the agitation and terror of a storm, they too frequently forget.

To add to the danger, a strong coastal crosscurrent runs between the two lines of bars. Survivors forced to leave their ships between the bars are helpless in this current, which may run more than three and a half miles per hour; they may experience the agony of seeing shore but being unable to get nearer to it.

One of the early victims of the outer shoals off Eastham-Wellfleet was the pirate ship *Whidah*, under command of Samual Bellamy. This ran aground in 1717. The sands covered some one hundred bodies of the crew, along with an unknown amount of pirate gold plunder. Early historians of the eighteenth century reported findings of old coins on the beach in this area; and piecemeal remains of the ship itself have been seen once or twice at very low tides. It was to plunder this wreck that a whaleboat took the risk of becoming the first large vessel to venture across the Cape by way of Jeremiah's Gutter.

The Indians had encountered these shoals in their days of sovereignty on Cape Cod. Probably they discovered them on their fishing trips. So impressed were they by the size and extent of the offshore bars that they assumed them to stretch from "the main at Pawmet to the Isle of Nauset" and thence "beyond their knowledge" into the sea. (The Isle of Nauset was said to lie about eight miles offshore opposite Nauset Light; it was visible in the seventeen hundreds but gone by 1875.)[18]

In 1602, Gosnold explored this region and found the surf-capped shoals of Chatham. Seventeenth-century French and Dutch fishermen, who probably knew this coast better than we realize, long knew the Chatham bars as *Mallebarre* (wicked bar). When the Pilgrims came and set sail southward along the Cape's eastern shore, aiming for the Hudson River and New Jersey, they were turned back by "dangerous shoals and roaring breakers." Swinging north-

[18] See p. 255.

ward instead, they anchored off what is now Provincetown and thence crossed Cape Cod Bay to Plymouth. Thus are the trappings of history often woven of the stuff of geology.

Peaked Hill Bars are the most northerly of the shoals along the Cape's outer shore. So dangerous are they that although they lie a half mile from land, the *Coast Pilot* warns all vessels not to approach the shore within two miles in this area. Fed by shore currents driven by southeast winds, Peaked Hill Bars are slowly growing, swelling with debris from the cliff-edged plain of Truro.

They claim most of their victims during northeast storms. "When a real nor'easter blows," writes Henry Beston, "howling landward through the winter night over a thousand miles of gray, tormented seas, all shipping off the Cape must pass the Cape or strand. In the darkness and scream of the storm, in the beat of the endless, icy, crystalline snow, rigging freezes, sails freeze and tear—of a sudden the long booming undertone of the surf sounds under the lee bow—a moment's drift, the feel of surf twisting the keel of the vessel, then the jarring, thundering crash and the upward drive of the bar."

In November 1778, a French ship chased the British man-of-war *Somerset* onto the southern continuation of these bars. The *Somerset*, made famous by Longfellow's *Paul Revere's Ride*, had long been stationed off the Cape, a mockery to the colonists. When it struck the bar, strong waves broke it apart, drowning many crew-men and sweeping in the wreckage with the tide. Jubilant Cape Codders took 480 prisoners and grabbed everything portable they could find, even parts of the broken wooden body of the ship. Shifting sands soon covered the wreck that remained, and have ex-posed it again from time to time. It was reported visible in 1880 and again in 1886. Even today, scarcely a summer goes by that someone does not think he sees revealed again the masts and spars of the phantom ship.

Today's power-driven ships can move confidently along the coast, no longer dependent on the vagaries of weather. Since 1798 there have been lighthouses on the Cape and, more recently, there is the Canal. Most important today, there is the Coast Guard, whose heroism is reinforced by technical training and specialized equipment for rescue work. Together, these have taken some of the old dread away from the shoals.

But the bars themselves are still there, fencing the shoreline like barbed wire and warning ships to keep their distance.

CHAPTER XIV

From Form to Form

. . . and the sands of Provinceland will be swept away as the oceanic curtain falls on this little one-act geographical drama.

WILLIAM MORRIS DAVIS
Physiographer, 1896

Just as the ancient origins of the Cape and Islands lie in the sea, so their futures too belong to the sea. Despite the ocean's never-ending land construction; despite all the relatively recent annexations of sand; despite the movement of spits and promontories from here to there, the Cape-Island area is smaller now than when these outlying rubbly extensions first appeared off New England.

When the ocean finally overcomes the last narrow strip of the glacial Cape and the last mounds of the ice-built Islands, it will have depleted the source of its building materials in that area. No longer will it fashion new land extensions and bedeck the old with them. It will play about its narrowing sandy spits, gnawing at them, causing them at last to disintegrate and to be drawn to the quiet sea floor. Who can know, however, where the currents of the next millennia will deposit the materials which now form the Cape and Islands?

Drowned Lands

Besides its destructive action, the sea in the past simply has flowed over parts of the land. Certain glacial lands already have vanished beneath the waters around Cape Cod and the Islands. These include the onetime land bridges between Cape and mainland and Cape and Islands, as we have seen. Such too is Georges Bank. Immediately after the final ice left, much of the area now occupied by the large bays and sounds was solid land, which stretched for many miles east of today's coasts. As the ice relinquished its water and sea level slowly rose, the lowlands became waterways and the highlands, islands. The outermost dots of land in this region came to be Georges Bank, the highest part of the eastward extension of the terminal moraine. Had our civilization followed more closely in time the wake of the retreating ice, Georges Bank might have ranked with Marthas Vineyard and Nantucket as a fishing port and summer resort island. The sea, however, continued its rise and covered this outlying moraine by a few feet of water, before the written memory of men.

Through the pages of history we see other onetime islands rising from the sea, partly obscured by the mists of time and legend.

Old charts drawn prior to the eighteenth century show an island lying about nine miles southeast of Chatham (*Figure 29*). Settlers named it Webb's Island. According to tradition, this was no mere shifting sandbar, but an established twenty-acre tract of good land. It was said to have been covered with sufficient red cedar to provide wood for Nantucketers to log; in fact, during the last century, stumps with ax marks on them still were drifting ashore at Chatham.

Some historians think the existence of Webb's Island only legendary. It seems strange indeed that Nantucketers would brave the dangerous shoals to get wood from it. Furthermore, on some seventeenth-century maps of the coast, Webb's Island is missing. Neither Champlain's 1606 chart nor the map drawn by Captain John Smith shows it. Although both Champlain and Smith were competent map makers, it is possible that they missed such a small wooded area. Its generally accepted location is near the shoals that they did their best to avoid.

If this miniature lost Atlantis did exist, the slow rise of the sea and the attack of waves has worn it down. By 1700, certainly, water covered it. Submerged, it may have served as a building block for the present Monomoy, for some old charts place it in about this position. By the middle of the eighteenth century, eighteen feet of water lay over the alleged site of Webb's Island. Today, all that remains is the clean sand of the Chatham shoals.

Another island which existed off Chatham as a somewhat more established fact was Ram Island (or Scotchpenacot or Cotchpinicuit, according to the Indians). This lay southeast of Allen Point, the headland east of North Chatham. While Ram Island existed,

Nauset Beach was discontinuous. It split just northeast of the island and bent landward to join North Chatham, thus giving protection to the island. Ram Island rose twenty feet from the water and had some thirteen acres, even a building of some sort on it. It was used as pastureland until 1851, when the Minot Gale whipped the sea over it and it was swallowed forever.

A northern counterpart to Webb's and Ram Islands was Nauset Island (or *Ile Nauset* as John Smith named it). It lay opposite Eastham. Leif Ericson in 1003 apparently encountered it, and the Indians knew it too. When Gosnold, in 1602, reported dangerous shoals in this vicinity over which the sea broke, he may have been speaking of this island, which was pictured as shrub-covered on Champlain's excellent 1605 map of Nauset Harbor. Later settlers gave the island the intriguing name of Slut's Bush. By the end of the nineteenth century, Nauset Island, like the others, had vanished in the sea.

In Cape Cod Bay an island once lay outside Wellfleet Harbor, just south of Jeremy Point, the southern tip of the sandbar which encloses the harbor. In fact, Billingsgate still exists as an island at low tide. It forms a nucleus for the extensive shoals which stretch southwest from Wellfleet into the Bay. There were cedars once upon Billingsgate Island, and it was on the submerged island that a fisherman claimed to Thoreau to have found water-covered cedar stumps "as big as cartwheels."

We can only speculate as to what other forested lands Cape Cod has lost. Together they must have added up to some sizable acreage. In 1841, off Yarmouth, Hitchcock found cedar stumps reaching over an area which extended more than three miles northward into the Bay. Around that time too, a layer of swamp soil with tree roots could be seen below the beach sand at low tide, west of Bass River at South Yarmouth. Beyond Witchmere Harbor, Centerville, there are submerged tree trunks below low tide. Off Provincetown, stumps of pine and cedar have been found eighteen to twenty feet deep, and fishermen have reported peat there, thirty feet below sea level.

Off Nantucket Island, where Tuckernuck and Muskeget appear today, there were once four substantial islands. Two since have dropped out of sight to become part of Tuckernuck Bank and the nearby shoals. Great Point, Nantucket's northern tip, probably was part of a large glacial island once. The sea's attack against its remaining sands still is reducing the size of Great Point by a rate as high as fifteen feet a year, and at the same time is tying its lingering core to Nantucket by a current-built sandspit.

The surrender to the sea of these old ice-built lands was due only partly to the attack of winds and waves. Primarily they vanished simply because the rising sea flooded them following glaciation, capturing them without a battle.

Increasing Inroads by the Sea

Other inroads of the sea are evident, as we have seen, in the channels which groove the southern part of the Islands' outwash plains, where the sea has drowned their mouths. And there is evidence too, of course, on the plains of the Lower Cape in the east-west valleys or "hollows," the heads of which once must have been in glacial debris farther east than the present coast. Pamet River's valley, where the Cape forearm begins to swing around to the west, has been beheaded by the sea, creating a tidal stream along the entire length of the channel. Farther south, tidal waters reach only partway up Wellfleet's Blackfish Creek channel, indicating that Blackfish Creek is probably in its present state what Pamet River was in its past. When the cliffs at McGuire Beach, South Wellfleet, recede to its head, Blackfish Creek will become a Pamet-type estuary.

The drownings by the sea indicate an erosion rate which certainly has increased throughout the postglacial centuries. For instance, immediately following ice retreat there was no sea erosion at all on today's shores. On the other hand, let us consider the probable scene several thousand years hence. If the earth's lingering ice sheets completely finish their melting, they will flood the earth's coastal areas with some 330 additional feet of water. This will drown whatever remains of the Cape and Islands at that time. This might be considered the ultimate stage of erosion.

Thus, from the end of the Pleistocene to a future end of the earth's glaciers, the Cape and Islands will have passed from no sea erosion of their shores to complete erosion. Our tiny moment of geological time places us somewhere between these extremes, witnesses to increasingly active erosion.

The Power of the Surf

At its present slow rise, the ocean will be able to overcome what remains of the bulky moraines and sloping plains of the Cape and Islands mostly by its unremitting wave-borne attack. Wind and waves today, as always, are in the midst of such an offensive. So unrelenting is the assault that it should level these headlands long before the world's remaining ice sheets melt and flood the lands.

The inroads of the sea are very real to those who have built homes atop high bluffs, within sight and sound of the surf that may someday exercise its right of eminent domain over their homesteads. When the wind piles water over the beach and against the cliffs, it imparts to it a fearful power. A wave ten feet high and a hundred feet long—a not unusually large wave—can exert a pressure of more than 1600 tons against every square foot it strikes. Storm waves may reach forty feet high and 500 feet long, and these crash against the opposing land with a pressure nearly four times that of a ten-foot wave. When such waves strike the cliffs, their pressure compresses the air within the pore spaces of cliff debris. This air struggles to free itself with explosive might, causing entire sections of cliff material to burst loose and fall.

To add to its power, the sea itself gets higher under these conditions, pushed up against the land so that the havoc of the waves may reach even farther inland. During the spectacular Minot Gale which hit this coast in 1851, the waters rose eighteen feet in Barnstable and seventeen feet along the Bay shore of the Lower Cape. They stopped just short of the fires in the Sandwich Glass Company furnaces on the Upper Cape. Years later, during another storm, Cape Cod Bay's water coursed over the Great Marshes and reached Barnstable's main highway (today's Route 6A).

In addition, whenever storm seas hit the land and pound against the cliffs they tear away the lower parts, scooping out the foundations of soils and trees and buildings on the highlands. And always there is the endless bombardment of wave-carried pebbles and sand, which help to eat away the weaker parts of the banks. When a cave has been hollowed out and the top of the cliff no longer can support itself, it collapses, yielding its substance to the surf.

Hurricanes

Sometimes still another form of attack is launched at the Cape and Islands by the tropical hurricane—although old-time Cape Codders may have called them simply "southwesters" or "northeasters" and taken them in their stride. When these infrequently reach the New England coast during the hurricane season from August through October, they come across many hundreds of miles of open sea. They announce their coming by a fall in barometers, a rise in the wind-driven sea, and a coupling of unusually high tides with rough storm waves. Flood damage from high tides may be severe in certain shore areas.

Storm waves begin to pound the shore with heavy cadence while the eye of the storm may be still as much as 500 miles away; and the beat will continue unbroken until the storm passes. Hurricane winds whirl in a counterclockwise vortex. As we face along the direction of hurricane movement, those winds which blow on our right side are the most destructive. When the storm itself strikes the land, it is these winds which may play havoc with everything exposed upon it and not secured. Propelling the surrounding water to tremendous power, they may in a few hours create an island, close off a bay, or steal someone's home or backyard. A Gay Head resident vividly recalled to the writer how, during a hurricane a few years ago, he watched more than fifty feet of the nearby cliffs, undermined, crash into the water.

Daily Wearing Away of the Land

The tearing down of the land, of course, is a product not only of the sudden orgies of tempests, but also of the slow nibbling of daily wind, rain and waves. Ordinary waves level the shoreline by a system of graduated taxation, filling the bays and planing the promontories, paring all to a featureless pattern. If you build a house on a headland or the point of a peninsula jutting out into the sea, you or your heirs will pay for a spectacular view and the percussion music of the surf with inevitable stoping away of your land. The breakers which roll to shore even in calm weather bend so as to concentrate their energy against any part of the shoreline which projects more than others.

When a wave reaches shallow water, as we have seen, it slows up as its motion scrapes bottom. Where there is a headland, that part of the incoming wave in front of the headland hits bottom first, for there the sea floor is shallower farther out. The wave slows there, while the rest of it continues in at full speed until it too feels bottom. In this way the wave bends around the headland, concentrating the strength of its attack against it from all possible directions.

In quiet coves the reverse happens. The wave spreads out and dissipates its energy, normally causing little if any erosion, and leading instead to sand deposition.

The even stretch of the outer shore of Cape Cod is an excellent example of the end result of these processes.

But water is not the only thing that wears back the land. Let us consider a small portion of the cliffs, whether at North Truro, at Gay Head, or at Tom Nevers Head. The wind which blasts these bluffs carries sand which pits their faces like a battery of Lilliputian machine guns, dislodging grains of sand and gravel. The loosened pebbles and sand grains tumble, roll, jump down the steep cliff face to its base. As they move, sometimes they set up miniature landslides, until the clatter of stones can be heard through the dominant roar of the sea. The freed sand gathers speed, tiny streams join together to form small rivulets of sand; these, like rivers of water, erode their channels and pick up more sand. Eventually a pile of loose debris or talus grows at the base of the cliff. The sand grains arrange themselves with a lower angle of repose than the clayey, semifirm cliff above them, which rises everywhere about thirty-five degrees from the beach. This slumped debris forms a sort of transition zone between the flat beach beneath and the crumbling cliff above. High waves will level the talus, but it quickly will form again, for the winds never cease.

Cape Cod Land Losses

Thus, structures raised by men may become helpless and indefensible. To Cape Codders and Islanders, such ravages are unwelcome but not unexpected, for inhabitants know well the power of gale winds and surf, and the long-term, inexorable march forward of the ocean.

In a United States Coast and Geodetic Survey report of 1889, H. L. Marindin said of Cape Cod's outer shore:

> Summing up the changes [from 1848 to 1888] along the entire length . . . [Chatham to North Truro], we find that 32,233,030 cubic yards of earth and sand have disappeared. This volume can best be understood by supposing it deposited on . . . the 55 acres included within the Capitol Grounds in Washington, D.C., which it would cover to a depth of 375 feet—in other words, the statue of Freedom on the dome of the Capitol would be buried to a depth of 67 feet.

If we walk up Nauset Beach from Nauset Light to Cape Cod Light, we find ourselves trending more and more to the west. By the time we reach Cape Cod Light we are walking more nearly west than north. Continuing along the beach, we end up following to the southwest an arching sandspit which ends at Race Point.

The arching of the land here is the result of northeast storm winds which drive waves to eat away at the outer shore relentlessly. Even from sea-built Provincetown the ocean has taken back part of what it gave, forcing the Peaked Hill Coast Guard station to move back twice. At Cape Cod Light, northeast gales drive the edge of the Truro plain westward at a rate measured at eight feet in some years. The three or four government acres where the lighthouse now stands are part of a ten-acre tract bought in 1787; the rest is lost to the sea.

These bold cliffs are the most vulnerable of all the Cape to storm attacks. Although the summertime prevailing winds of this region are from the southwest, the sea puts on a different face for winter. Powerful winter storm winds sweep across the face of the open Atlantic from the northeast or, sometimes, southeast, and lash the coast.

Hidden but effective, Georges Bank and the Nantucket Shoals break some of the force of southeast waves. But as Ralph Waldo Emerson said of Nantucket, "They say here that a northeaster never dies in debt to a southwester, but pays all back with interest." Northeasters have a free running start across nearly 300 miles of water if they come by way of Nova Scotia, even more if they bypass that corner of land by traveling over the Bay of Fundy. These waves are effective erosive agents. After a storm we can stand on the beach and watch the swelling, churning breakers sweep toward us. They are opaque and brown, not the usual green–brown with the torn and powdered substance of the land.

The average yearly rate of coast recession between Cape Cod Light and Nauset Light, measured over a period of forty years, is nearly three and a half feet. This represents the removal of more than thirty million cubic yards of material—54,000 cubic yards along each mile of shore.

Nauset Light in Eastham stands on the edge of the tablelands. South of the lighthouse, this plain and its sea cliffs drop down to the knobby Eastham kame fields, which have been flattened at the base of the lighthouse hill to form a town parking area. At the present writing there is a snack stand on the northern edge of the parking field, with picnic tables between it and the nearby edge of the cliff. Across the steep cliff face beyond lies a thick, tangled layer of dead brush. This is not "a shameful dumping ground for trash," as the writer overheard a summer visitor say; it is a barricade, an effort to prolong the life of the picnic area and the snack stand itself. A few years ago there was another picnic area comfortably fenced in at a point where the brush pile now covers the loose debris of the crumbling bank. In a couple of years the seaward end of the fence was gone. Later the stand itself stood balanced, like a diver about to plunge, on the cliff edge. The building was moved back a few years ago to its present, no less temporary position.

Land has vanished from beneath the Nauset Lighthouse four times, each time necessitating the construction of one or more new ones farther back. Until quite recently there was visible halfway down the cliff face the round brick foundation of one of the wooden lighthouses which stood in triplicate at Nauset before the present light came in 1923. Those three lighthouses were rescued to become parts of nearby summer homes, while their foundations went to the sea.

And the sea endlessly harasses the asphalt parking lot beside the lighthouse. A single good winter gale can undermine its entire front edge and send benches, foundations, concrete posts and chunks of asphalt tumbling down the cliff. The ladder-like wooden steps down the face of the bluffs must be removed winters, "otherwise," says Eastham selectman Maurice Wiley, "eight years out of ten they would be lost" (*Figure 45*).

The highest rate of erosion here measured shows that this part of the coast takes only slightly less of a beating than that at Truro,

Figure 45. *After a storm. This havoc was caused by a single storm near Nauset Light.* CAPE COD PHOTOS

losing up to five feet in some years. Chatham, since 1880, also has lost land at an average of five feet per year; Chatham Light has taken three large steps backward.

Cape Cod's Vineyard and Nantucket Sound shores are more sheltered. Southeasters, however, can sweep across these waters and take large bites from the land. Many a broken seawall attests to their power. A northwest wind likewise can turn Cape Cod Bay into a small ocean which may make significant inroads into the morainal bluffs.

Cliff dwellers along the Bay have tried covering their banks to protect them, as the cliffs at Nauset Light are covered. Rocks, dead trees, brush, concrete—anything is tried to keep the material of the bluffs from an unhealthy exposure to storm waves and to allow plants to root themselves behind and within the buffer zone. This slows up, but hardly checks, the sea's attack.

More concentrated efforts at control may get underway before long, if enough money can be raised. A U. S. Army Corps of Engineers study indicates that erosion on Eastham's bayside could be nearly stopped with nine 300-foot, 1400-ton rock groins a thousand feet apart. The cost of such a project, at this writing, is estimated at up to $250,000, to which the federal government, the state, the county and the town would be expected to contribute.

Marthas Vineyard Land Losses

Marthas Vineyard and Nantucket are no less vulnerable. On the Vineyard, at the northeastern tip of the sandy spit known as Cape Poge, the cliff beneath the lighthouse loses a record ten and a half feet each year to northeasters. These storms sweep into Vineyard Haven Harbor and against Oak Bluffs with destructive force. A gale in 1898 sank or drove ashore fifty ships along Beach Road.

Along the east side of West Chop one hundred years ago, boats of fishermen could find haven in a small inlet. The sea since has removed it completely. Meanwhile, the same attacking waves have forced back the West Chop Lighthouse once or perhaps twice since it first was built.

On the Vineyard's southern coast, next-door neighbors Nashaquitsa and Gay Head show the effects of erosion on differing substances. "The former [glacial] is falling rapidly away, while the latter [sea-laid sediments] has been kindly dealt with. One is treated as an intruder upon the ocean's domain; the other, as a peaceful settler," wrote an early geologist. Southward-facing Nashaquitsa, opposed to the open sea, loses three to five and a half feet of its shore each year. But on the southwest part of the Vineyard, the last remnants of firm New England coastal-plain clays armor Gay Head against such rapid destruction. Besides, this neck of land is partially shielded against the brunt of storms by the land close by on its north. Gay Head, therefore, loses only about a foot or two of land in normal years.

Even this usual slow rate has its inevitable effect. The first lighthouse at Gay Head had to be replaced in 1858 by the present one. Furthermore, the treacherous shoal which stretches westward from the Gay Head cliffs, Devils Bridge, is a hidden string of boulders; this indicates that the shoal may be built of stolen cliff materials and that its extent—at least a mile northwest of the land—may represent the original reaches of this part of the island.

Nantucket Land Losses

"On the seashore at Nantucket," said Ralph Waldo Emerson in 1847, "I saw the play of the Atlantic with the coast. Here was wealth; every wave reached a quarter of a mile along shore as it broke. . . . Ah, what freedom and grace and beauty with all this might! . . . Place of winds, bleak, shelterless, and when it blows, a large part of the island is suspended in the air and comes into your face and eyes as if it was glad to see you."

Responding to such winds and their driven seas, Nantucket Island already has shrunk by many hundreds of acres along its southern, eastern and western sides. Its southern coast loses land at the rate of nine acres each year. Near Miacomet Pond west of Surfside, the sea takes as much as eight feet a year; and in one fourteen-year period (1889–1903) it averaged more than seventeen feet a year.

In the first half of the last century there was a resort hotel here, the *Surfside*. A railroad ran in front of the building. One day storm seas reached the tracks, smothering them with sand. The tracks were moved back and again the surf found them. Once more they retreated and once more were destroyed. Eventually the railroad was rebuilt across central Nantucket. But the sea, not to be thwarted, half consumed the hotel itself in 1916.

East of Miacomet, the jog in the coastline at Surfside, which we can see on a map, is the result of wave erosion on the east and the simultaneous building by currents on the west of bars which close off outwash-plain ponds. For the triangular shoal known as Miacomet Rip, just southeast of the pond, breaks the power of incoming waves and channels their energy into the creation of current deposits.

As the attacking sea grinds its way into the waveward side of such bars and dumps the debris on the lee, the long ponds continually shorten. For example, we can see on a map that marsh-filled Weeweeder Ponds have a V-shaped pattern, unusual for a meltwater stream pattern. But if the V of these two ponds were extended into a Y, it would take on the more plausible pattern of a pair of tributary streams meeting and joining partway to the sea to form one stream. The V which remains today is simply the upper fork of this system, the rest having gone to sea.

We can, in fact, read a great deal concerning Nantucket's original extent and its rate of land loss by noticing on a map certain of the features which decorate the outwash plain. Comparing the ponds with their counterparts on Marthas Vineyard, we find the Vineyard's southern coast in a less advanced stage of erosion than that of far-flung Nantucket (*Figure 46*). The relations of shoreline and ponds on the Vineyard can be taken to indicate the situation on Nantucket at some time past.

Nantucket's Nobadeer Valley, which runs south from the airport, probably joined Madequecham Valley once, somewhere beyond the present coast, just as similar channels join today on the Vineyard. Let us extend the channels of these two Nantucket ponds hypothetically southward, following their present trends. We see that they could not have come together less than a mile and a half south of the present shoreline. So a mile and a half at least is probably gone from Nantucket.

Hints of things past can lie in names too. A long channel, one and a half miles east of Madequecham Valley, is called Forked Pond Valley. In historic times, it must have resembled today's Hummock Pond farther west, although its branches since have been erased by the sea.

Where the coast swings to the north and northwest, it yields to the force of winter gales from the north. Here six to ten feet of land vanish each year. North of Wauwinet, for instance, there is a thin strip of land named Haulover Beach. Its name goes back to the days when dories could be hauled across the narrow sand from the sea into the harbor, and so avoid the journey up and around Great Point. Sometimes sailors were saved the portage, however, and could bring boats afloat through the shortcut; for during heavy storms such as the 1896–1897 gales, the sea has pounded over the sand neck to create a shifting cut-through, Haulover Break. The sand derived from such attacks piles up south of Haulover on both

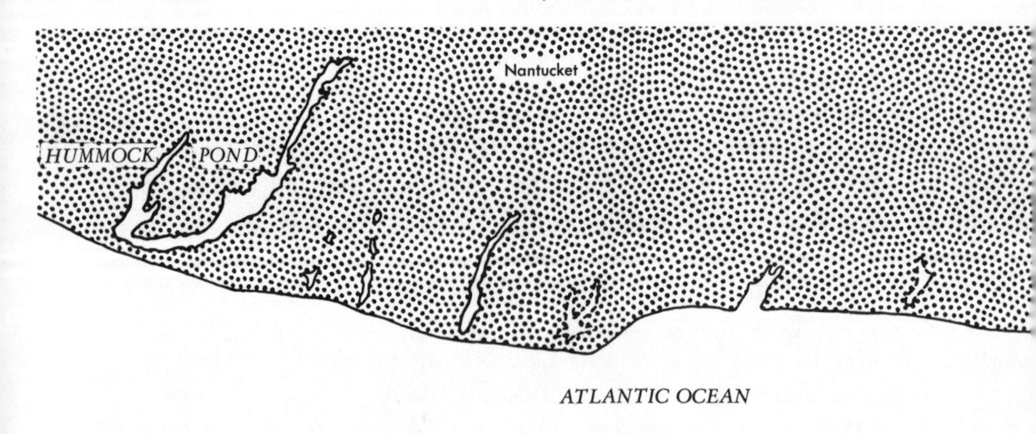

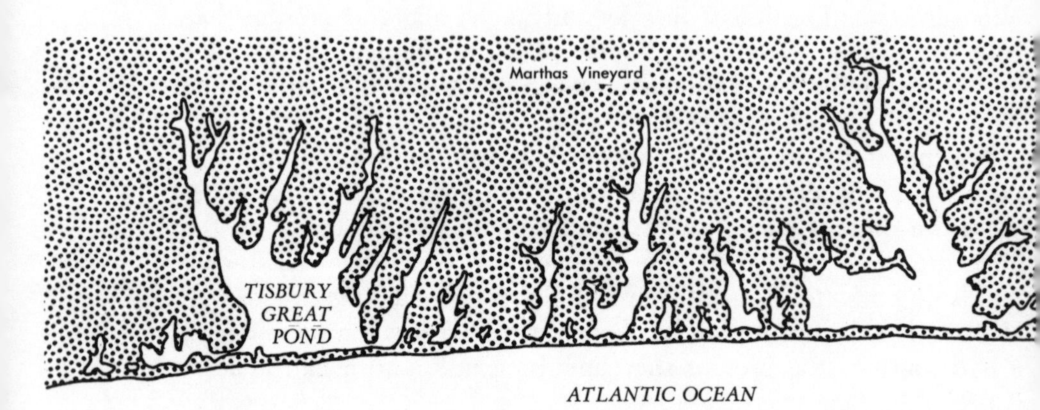

Figure 46. *The progress of erosion. The sea has eaten farther back into Nantucket's southern shore than into the analogous southern shore of Marthas Vineyard.*

ocean and harbor sides, while the sea turns its attack to the land about Nantucket Harbor.

On the east we can reconstruct the island's outwash plain to its original extent by continuing its gently sloping surface seaward beyond its wave-truncated bluffs. We find that if it continued it would reach sea level a mile south of the present beach; hence, about a mile of land is gone. Farther north, storm seas sometimes completely cut off Great Point from Nantucket.

Erosion at Tom Nevers Head is the most rapid of all. Here the coast swings northeast. Easterly storms slam against the headland with unmoderated strength.

On the other side of the island, we know that Tuckernuck Island was connected to Nantucket late in the eighteenth century, for farmers drove their young cattle to the island to feed across a low-tide land bridge every springtime. Today, only occasional storms transiently throw up such a beach connection.

Within a single week in 1961, the sea, driven by a hurricane, devoured 450 feet of the land at Maddaket. A group of beach homes had to be moved to safety. During a few months of the same year, Smith Point became an island. It is separated ever increasingly from Maddaket by a channel, created a quarter of a mile across by one hurricane and widened to a half mile by another.

In general, Nantucket's southeast coast loses less land each year than the southwest, for the former is partly buffered by the Nantucket Shoals. Crevecoeur gave them their due: ". . . these are the bulwarks which so powerfully defend this island from the impulse of the mighty ocean, and repel the force of its waves; which, but for the accumulated barriers, would ere now have dissolved its foundations and torn it to pieces."

How Erosion Is Hastened

To anyone who has seen such processes at work on the Cape and Islands, their ultimate result surely would seem to be the disappearance of these lands as we know them. And beyond these processes there is evidence, as we have seen, that the rate of erosion actually is increasing; sea level has been rising since ice retreat, and even if there were no active wave erosion, most of the Cape and Islands would become drowned should all remaining glaciers melt.

Besides the effects of the encroaching sea—which has risen at a rate of three and a half feet per thousand years between 1893 and 1930, and increased to nearly two feet per century in the next score of years—the shape of the land itself aids in hastening erosion further.

Even if sea level should become stabilized and stop rising, the erosion rate still would increase. Where land slopes away from the sea, erosion speeds up as it proceeds. The Cape's plains on its outer arm, for instance, slope westward from the eastern sea cliffs.

Each time waves excavate the lower part of these seaward bluffs, causing land to tumble down, the sea must rid itself of the debris before it can get back to its assault. Since the land slopes away from the sea, each time a piece is taken the remaining bluff is lower. When waves cause this lower bluff to collapse, they receive from it somewhat less debris to slow up their next attack.

In addition, where the land is narrow it is likely to be cut through comparatively rapidly, leaving the wider regions more completely surrounded by water. This water will develop longshore currents which skirt the land, removing bits of it more rapidly than was possible before the cut-through.

Forecast of an End

The glacial Cape once reached eastward perhaps two miles beyond the present eastern shore. The southern shores of Marthas Vineyard and Nantucket, as we have seen, once may have reached southward an extra mile or even two. In the 10,000 years since the glaciers left, these lands averaged a foot's loss each year. Today we can measure the erosion rate as three and a half feet a year. And as we have seen, it looks as if the loss will be even more rapid in the future.

Considering land losses along exposed shores and taking into account these various factors, Douglas Johnson has estimated that most of the Cape as we know it (and his arguments can be applied to the Islands as well)—the open plains, rolling meadows, pine forests, moors and dunes—will be nearly gone by the year 4000, totally gone 2000 years later. If erosion should cease to accelerate and should remain at its present rate, we may expect only 2000 additional years of grace.

The Cape forearm today is about 10,000 feet wide at Cape Cod Light; it widens to a maximum of 25,000 feet north of Wellfleet, and narrows again to 10,000 feet south of Wellfleet. At today's erosion rate, three and a half feet per year, the sea should break through the narrower parts in some 3000 years, and Provincetown and Wellfleet will become islands. It then would be a matter of only another three to five thousand years before even the widest parts would go. Cotton Mather once said that the name, Cape Cod, would cling to the peninsula "till the shoals of codfish be seen swimming on the highest hills"—an unintended prophecy.

Hint of a Beginning

Yet the picture is not entirely one of destruction. What will the sea do with all the debris it is taking from the present land? Let us turn aside from the Cape's Outer Beach with its eroding surf. If we walk instead over to the Bay—or up to Provincetown, or go to Nantucket's Sankaty beach—we shall see the other side of nature's give-and-take coin.

In some places the sea acts as a brake on its own erosive activities. High Head, the northern tip of the glacial Cape, is a good example of this. As we have seen, where waves once drove back the land there is now a quiet lagoon and protecting sand dunes, built of the debris which the sea acquired as a result of its robberies.

On Nantucket, too, there are cliffs which once plunged directly into the sea, those at Sankaty Head and Siasconset. The wide beach which since has spread itself out at their feet runs interference now against the attacking waves, so that they manage to bite into the bluffs only during high storms. Siasconset's beach advanced nearly a quarter of a mile between 1846 and 1890.

Furthermore, as wave-cutting everywhere provides the water with more and more debris, that which drifts out beyond the pull of shore currents settles. It creates a wide offshore platform. The more the sea erodes, the more this widens. Waves must drag their heels across it in order to reach shore. In so doing, they wear themselves out, and this slows up the rate of erosion of the land.

Not only does the sea cause a deceleration of its own destructive acts, but its land-building activities meanwhile are lively too. The preglacial sediments of Marthas Vineyard, torn from ancient lands to their west; the moraines and outwash plains making up most of the Cape and Islands, stolen from the north, are being recalled by the sea, it is true—but much will be used in the creation of new lands.

The Provincetown spit on the whole is lengthening rapidly. Despite the whipping it receives from the storms, Provincetown's Outer Beach grows wider yearly. According to Dr. John M. Zeigler of the Woods Hole Oceanographic Institute, the beach at Provincetown currently is increasing in width faster than Nauset Beach to its south is receding.

Cape Cod Bay becomes shallower yearly with blankets of mud derived from the eastern highlands. Thoreau noted among Cape Codders a belief which still exists in places, that the Cape is simply rolling over on itself, abuilding on the west as it loses on the east. The vast shoals between the Cape and Islands are continually stirring and growing, providing new hazards to boats.

Thus, what is taken away may appear in another place and form. In the far-off future, when, estimates indicate, the ocean will have cut completely through the Cape Cod, Marthas Vineyard and Nantucket that we know, our descendants even 150 generations hence may be walking beaches, climbing dunes, bathing in quiet bays of new capes and islands, born of the substance of these fragile outposts.

Appendices

APPENDIX A

GEOLOGIC TIME SCALE

Showing the Ages of Cape-Island Deposits and Their Relations to other Events

Era	*Period*	*Epoch(s)*	*Age (millions of years ago)*	*Important Events in Cape-Island Region*	*Important Events Elsewhere*	*Life*
Cryptozoic			3310 to 500	hidden bedrock; no visible record	formation of lowest rocks of Grand Canyon	worms; jellyfish; seaweeds
Paleozoic	Cambrian		600 to 500		formation of some of rocks of Appalachian Mountains	marine invertebrates; seaweeds
	Ordovician		500 to 440		Vermont marble; Pennsylvania slate	marine invertebrates; seaweeds; primitive fishes
	Silurian		440 to 400		rimrock at Niagara Falls; Michigan salt beds	marine creatures; primitive fishes; land plants; air-breathing insects
	Devonian		400 to 350		Catskill Mountain beds	fishes; amphibians; reptiles
	Mississippian		350 to 300		Mammoth Cave limestone	sharks; amphibians; reptiles
	Pennsylvanian		300 to 270		Pennsylvania coal deposits	forests; giant insects; spiders
	Permian		270		Appalachian	fern-like plants;

					Petrified Forest, Arizona	
	Jurassic		180 to 135		formation of Zion Nat'l. Park sandstone; Arches Nat'l. Monument	conifer trees; early flowering trees; early mammals
	Cretaceous		135 to 70	clays and lignite on Marthas Vineyard	chalk in Europe; uplift of Rockies	flowering plants; last dinosaurs; early mammals
Cenozoic	Tertiary (early)	Paleocene Eocene Oligocene	70 to 30	sediments beneath Provincetown	formation of Bryce Canyon beds; formation of South Dakota Badlands beds	mammals
	Tertiary (late)	Miocene Pliocene	30 to 600,000 yrs.	greensands on Marthas Vineyard and Georges Bank	birth of Mt. Rainier and Mt. Shasta; origin of valuable Gulf Coast oil fields	mammals
	Quaternary	Pleistocene	600,000 yrs. to 10,000 yrs.	Aquinnah Conglomerate; early tills; Gardiners Clay (Marthas Vineyard and Cape Cod); Sankaty Sand (Nantucket); moraines; outwash plains	widespread glaciation	early man
		Recent	10,000 yrs. to present	beaches; dunes; sandspits; bars; harbors; vegetation	continual changes of the face of the earth	modern man

Appendix B

IDENTIFYING CAPE-ISLAND MINERALS*

The tables which follow contain minerals most likely to be found on the Cape and Islands. They are arranged in order of increasing HARDNESS (Column B). HARDNESS is the ability of one substance to scratch another; thus, any substance can be scratched by a different substance of greater hardness. To test minerals roughly for hardness we use a few simple tools. These, in order of *increasing* hardness, are:

a fingernail – a copper penny – a penknife – a piece of glass

Each of these will scratch all before it on the list and, conversely, each will be scratched by all after it.

Each mineral has a definite hardness that can be related to the above tools. For instance, gypsum (#3 on the table) can be scratched by a fingernail. Calcite (#11) is harder than a fingernail and cannot be scratched by one. But calcite has the same hardness as a penny (when two substances have the same hardness they will scratch one another). So calcite will scratch a penny and gypsum will not. Thus, gypsum and calcite, which may look somewhat alike, can be distinguished on the basis of hardness. Once the hardness is roughly established, the other characteristics listed in Columns C to J should help to pin down the mineral in question.

Column F lists the places each mineral is most likely to be found on the Cape and Islands. As a cross-reference for Column F, there follows here the major mineral occurrences in this region and the minerals most commonly found in each (numbers refer to mineral numbers on the charts). It should be stressed that this is only a *general* list, and minerals not listed as *common* in one locality may still show up there.

Minerals Found as Parts of *Drift Rocks and Beach Pebbles*

1. talc
2. graphite
6. mica
7. chlorite
11. calcite
12. serpentine
14. hornblende
15. pyroxene
16. magnetite
17. hematite
18. feldspars
19. pyrite
21. epidote
22. olivine
23. garnet
24. quartz
25. chalcedony
26. tourmaline
27. beryl
28. corundum

* Two excellent field books which describe and picture common minerals are Pough's Peterson Field Guide and the Golden Nature Guide, both listed in the Bibliography, Part II.

Minerals Found as *Grains of Beach Sand*

13. ilmenite
14. hornblende
15. pyroxene
16. magnetite
18. feldspars
22. olivine
23. garnet
24. quartz
26. tourmaline
27. beryl
28. corundum

Minerals Found in *Cretaceous Clay and Lignite*

3. gypsum
5. kaolin
19. pyrite
20. marcasite

Minerals Found in *Tertiary Greensands*

4. glauconite
9. collophane

Minerals *Filling Cracks or Lining Surfaces of Rocks*

8. halite
11. calcite
23. garnet
24. quartz
25. chalcedony

Minerals Found as *Coloring Matter in Sands and Soils*

10. limonite
17. hematite

Mineral Found as *Bog Iron Ore*

10. limonite

Mineral Found as *Seashells*

11. calcite

Mineral Found Dissolved in *Seawater*

8. halite

A *Name*	B *Hardness*	C *Color*	D *Surface Appearance*	E *Cleavage or Manner of Breaking*	F *Where Fo in This Re*
1 talc	very soft, will mark cloth	light green; gray; white	pearly; greasy	perfect cleavage in 1 direction; splits into flakes	as intrinsic of *drift roc* and *beach bles,* from teration of minerals; igneous roc metamorphi rocks
2 graphite	very soft, will mark paper	black	metallic	good cleavage in 1 direction but seen only in large masses	as intrinsic of *drift roc* and *beach bles*
3 gypsum	softer than, to equal to, fingernail	colorless; white; pink	glassy	good cleavage in 4 directions	as crystals i *Cretaceous nites* of Ma Vineyard
4 glauconite	softer than, to equal to, fingernail	green when fresh; oxidizes to red on exposure to air	dull, earthy	breaks irregularly	makes up *T ary greensa* (q.v.) on Marthas Vi yard
5 kaolin	softer than, to equal to, fingernail	white; tan; red	dull, earthy	powdery	makes up p of *Cretaceo clay* of Mar Vineyard
6 mica	nearly equal to fingernail	black; white; tan	glittering; glassy	highly perfect cleavage in 1 direction, forming paper-thin flakes	a. as intrins part of *d rocks* and *beach pe bles:* igne rocks, me morphic r sedimenta rocks b. as grains *beach san*
7 chlorite	nearly equal to fingernail	green	pearly; glassy	perfect cleavage in 1 direction; splits into flakes	as intrinsic of *drift roc* and *beach bles,* from a teration of minerals: igneous rock metamorphi rocks, sedim tary rocks

G	H	I	J
ɔrm and Shape	*Other Properties*	*Chemical Composition*	*Remarks*
ıgs; shapeless es		magnesium silicate with water $Mg_3(Si_4O_{10})(OH)_2$	softer than chlorite (q.v.), and lighter-colored with greasier feel; when talc makes up an entire rock it is soapstone
streaks		carbon C	
colorless flaky s and needles		calcium sulfate with water $CaSO_4 \cdot 2H_2O$	will not fizz in vinegar as will similar-looking calcite (q.v.); softer than calcite
r green sand-sized		potassium, magnesium, iron, aluminum silicate with water $K(Fe,Mg,Al)_2(Si_4O_{10})(OH)_2$	once used for fertilizer
ɔf clayey sediments		aluminum silicate with water $Al_4(Si_4O_{10})(OH)_8$	can be molded when wet, hardens when dry; may be fired into ceramics
n flakes, single or ıped	a. easily flaked off rock with knife point	a family of potassium, magnesium, iron, aluminum silicates with water $KAl_2(AlSi_3O_{10})(OH)_2$ and $K(Mg,Fe)_3(AlSi_3O_{10})(OH)_2$	important rock mineral; flakes of the tan variety may look like gold glittering in a rock
n flakes which ɛe the sand rkle			
less coatings or -surfaced masses	looks like green mica	magnesium silicate with water $Mg_3(Si_4O_{10})(OH)_2Mg_3(OH)_6$	

A *Name*	B *Hardness*	C *Color*	D *Surface Appearance*	E *Manner of Cleavage or Breaking*	F *Where Fo in This Re*
8 halite	nearly equal to fingernail	white	glassy	cleaves into cubes	from evap tion of *sea water;* line lows in ro exposed be tween tide
9 collophane	soft, with variable hardness	white; tan; brown	dull	pulverent	in nodules *Tertiary se ments* of M thas Viney
10 limonite	powdery form makes it seem soft	brown to yellow	earthy	pulverent	a. as brow yellow c ing matt *preglacia glacial s ments* ar soils b. as bog ore
11 calcite	same as penny	any color	glassy; dull	perfect cleavage in 2 directions at 75° angle	a. as *entir rocks* (li stone; m b. filling *c in rocks* c. as *seash*
12 serpentine	same as penny	light or dark green	greasy; waxy	breaks irregularly	as intrinsi of *drift ro* and *beach bles,* from teration of other min igneous ro metamorp rocks
13 ilmenite	equal to knife	iron black; may be coated with tan iron oxide	metallic	breaks irregularly	as grains *beach san*
14 hornblende	equal to glass	dark green to black; sometimes with tan stain	glassy; dull	perfect cleavage in 2 directions, 120° apart, causing pattern of diamond-shaped cracks	a. as intri part of *rocks* an *beach p bles:* ig rocks, metamor rocks b. as grai *beach s*

G *orm and Shape*	H *Other Properties*	I *Chemical Composition*	J *Remarks*
s or granular ers	tastes salty	sodium chloride $NaCl$	common table salt; rock salt
ular, lumpy masses	smell rotten when broken	calcium phosphate with water and various impurities	collophane is general term for phosphatic material; no particular mineral
owder or tarnish		iron oxide with water $FeO(OH) \cdot nH_2O$	a. common alteration product of magnetite, hematite, pyrite (q.v.)
ongy masses			
assive; dull, dirty mestone); sparkling ystals (marble) hite, dogtooth-aped crystals or id white masses	fizzes when scratched and moistened with vinegar or other acid	calcium carbonate $CaCO_3$	a. easily destroyed in ice transport, therefore not common among drift rocks
eless coatings or es		magnesium silicate with water $Mg_6(Si_4O_{10})(OH)_8$	
, rounded grains	weakly attracted to magnet	iron titanium oxide $FeTiO_3$	once common with magnetite on some beaches of Marthas Vineyard; can be distinguished from magnetite (q.v.) sand only chemically
ongated prisms or edles		complex silicate of calcium, iron, magnesium, sodium, aluminum, with water	important rock mineral; may resemble "petrified blades of grass"; (see tourmaline, ※26)
rk, long thin grains			

A *Name*	B *Hardness*	C *Color*	D *Surface Appearance*	E *Cleavage or Manner of Breaking*	F *Where Fo* *in This Re*
15 pyroxene	equal to glass	many colors; usually dark green to black, sometimes with tan stain	dull; glassy	perfect cleavage in 2 directions at right angles	a. as intrin part of *rocks* an *beach pebbles:* igneous metamor rocks b. as grain *beach sa*
16 magnetite	equal to glass	iron black; may be coated with tan iron oxide	metallic	usually breaks irregularly	a. as intrir though grained, of *drift* and *bea pebbles* b. as grain *beach sa*
17 hematite	equal to glass	red-brown; black	metallic; dull	breaks irregularly	a. as intrir part of *rocks* an *beach pe bles:* ign rocks, m morphic sediment rocks; al the ceme and grai coatings *sand* and *sediment* b. as *beach pebbles*
18 feldspars	slightly harder than glass	white; tan; pink; gray	glassy; dull	perfect cleavage in 2 directions at right angles; yields blocky fragments	a. as intrin part of *rocks* an *beach pe bles:* ign rocks, metamor rocks, sediment rocks b. alone as *beach pe bles* c. as grain *beach sa*

G *'orm and Shape*	H *Other Properties*	I *Chemical Composition*	J *Remarks*
ıbby prisms or ınular		complex silicate of calcium, iron, magnesium, sodium, aluminum	important rock mineral; a major component of the peridotite (q.v.) in early till on Marthas Vineyard, derived from Rhode Island
ırk, shapeless or ngated grains			
:tahedrons (double ramids) or 12-faced ids	strongly attracted to magnet	iron oxide Fe_3O_4	a. usually too disseminated through rock to be noticed; crush rock and magnet will pick up magnetite
lack, rounded grains			b. once common on some beaches of Marthas Vineyard
ains or irregular ısses		iron oxide Fe_2O_3	a. colors many preglacial sediments on Marthas Vineyard red and stains much glacial drift on Cape Cod and the Islands
·d lumpy masses			b. red ocher; when hollow known as Indian paint pots
apeless or rectangu- masses	sometimes has minute parallel lines across surface	potassium, sodium, calcium aluminum silicates $KAlSi_3O_8$ to $NaAlSi_3O_8$ to $CaAl_2Si_2O_8$	important family of rock minerals; in a rock with quartz, feldspar will be less glassy, more dull, and will show cleavage, whereas quartz will not
usually retain blocky ıpe, edges and rners rounded			

A *Name*	B *Hardness*	C *Color*	D *Surface Appearance*	E *Cleavage or Manner of Breaking*	F *Where F in This R*
19 pyrite	slightly harder than glass	brass-yellow	metallic	breaks irregular-ly	as intrinsi of *drift ro* and *beach bles:* igneous ro metamorph rocks, sedimentar rocks
20 marcasite	slightly harder than glass	bronze-yellow to whitish, often coated with tan	metallic	breaks irregular-ly	as crystal masses in *Cretaceous nites* of M Vineyard
21 epidote	harder than glass	pistachio-green	glassy	good cleavage in 1 direction	as intrinsi of *drift ro* and *beach bles:* igneous ro metamorph rocks
22 olivine	harder than glass	olive-green	glassy	breaks with jagged edges like glass	a. as intri part of *rocks* an *beach pe bles:* igneous metamor rocks, sediment rocks b. as grain *beach sa*
23 garnet	harder than glass	red; brown; black	glassy	breaks irregular-ly	a. as intrin part of, crystals the surfa of, *drift* and *bea pebbles:* igneous metamor rocks, sediment rocks b. as grain *beach sa*

G *Form and Shape*	H *Other Properties*	I *Chemical Composition*	J *Remarks*
›es or 12-sided solids; ›nular masses		iron sulfide FeS_2	fool's gold
›ules of metallic ›terials		iron sulfide FeS_2	same chemical composition as pyrite but different atomic structure; color different
›nular or fibrous ›sses		calcium, aluminum, iron silicate with water $Ca_2(AlFe)Al_2O(SiO_4)$-$(Si_2O_7)(OH)$	harder than and different green than serpentine and chlorite; less common than hornblende and different color
›granular masses or ›attered grains	rarely found in same rock as quartz	magnesium iron silicates $(Mg,Fe)_2SiO_4$	with pyroxene, makes up the peridotite in the early till of Marthas Vineyard, derived from Rhode Island
›ounded green glassy ›ains			
›early spherical ›ains or clusters of ›ains	rather heavy	magnesium, manganese, iron, calcium, aluminum silicates	important gemstone when deep, clear red
›mall spherical red ›ains			

A *Name*	B *Hardness*	C *Color*	D *Surface Appearance*	E *Manner of Cleavage or Breaking*	F *Where Founc in This Regio*
24 quartz a. ordinary	harder than glass	a. colorless; white; gray	shiny; glassy; greasy; smooth; rough	breaks with jagged edges like glass	a1. as intrinsi part of *drif rocks* and *beach peb- bles:* igneou rocks, metamorphi rocks, sedimentary rocks a2. as grains *beach sand, till, and pre glacial sedi- ments*
b. amethyst		b. purple			b1. alone as *beach pebb* b2. as grains *beach sand*
c. milky quartz		c. milk-white			c1. alone as *beach pebb* c2. as grains *beach sand* c3. as *veins* f ing cracks i rocks
d. rose quartz		d. pink			d1. alone as *beach pebb* d2. as grains *beach sand*
e. rock crystal		e. clear, color- less			e. along *crack* or in *hollow* in rocks, es pecially *dri rocks*
25 chalcedony a. agate	harder than glass	a. banded or variegated, mainly gray; red; brown	waxy; glassy	breaks with jagged edges like glass	a. alone as *beach pebb* and in *drift rocks*
b. jasper		b. red			b. alone as *beach pebb*
c. flint, chert		c. gray or black			c. alone as *beach pebb*

G *Form and Shape*	H *Other Properties*	I *Chemical Composition*	J *Remarks*
shapeless glassy rains, sometimes ched into raised lief on rock surface		silicon dioxide (silica) SiO_2	a1. important, nearly ubiquitous rock mineral
rounded white or lorless grains			a2. makes up 90% of sand in this region
rounded purple ebbles with smoothly orn edges			b1. once reported common at Maddaket, Nantucket
rounded purple ains			
smooth or cavernous hite masses, surfaces unded			c1. common among beach pebbles
rounded white aque grains			
solid white bands or ripes			
rounded pink bbles			
rounded pink grains			
ng pointed crystals ith 6-sided cross-ctions, in cracks in cks			e. sometimes found by splitting rocks open
unded masses; ugh surfaces	a. color bands not clear on surface; must be broken open	silicon dioxide (silica) SiO_2 (minutely crystalline quartz)	a. on all beaches but very common on SW Cape Cod and Marthas Vineyard
d, compact hard asses, rounded ooth			b. on all beaches
ard, rounded com-ct masses			c. used by Indians for arrowheads

A *Name*	B *Hardness*	C *Color*	D *Surface Appearance*	E *Manner of Cleavage or Breaking*	F *Where Foun[d] in This Regio[n]*
26 tourmaline	harder than glass	usually black; may be brown; red; green	glassy	breaks with jagged edges	a. as intrinsic part of *drift rocks* and *beach pebbles:* igneou[s] rocks, metamorphi[c] rocks b. as grains o[f] *beach sand*
27 beryl	harder than quartz	blue-green; yellow	glassy	cleaves imperfectly	a. as intrinsic part of *drif[t] rocks* and *beach pebbles:* pe[g]matite dike[s], metamorphi[c] rocks b. as grains [of] *beach sand*
28 corundum	harder than quartz (softer than diamond)	brown to red	glassy	breaks in blocky fashion	as intrinsic though minor part of *drift rocks* and *be[ach] pebbles:* igne[ous] rocks

G *Form and Shape*	H *Other Properties*	I *Chemical Composition*	J *Remarks*
.g, thin, shiny prisms :h lines parallel to ·ir lengths; in cross- ·tion these form ıerical triangles		a complex boro-aluminum silicate with sodium, calcium, iron, lithium, magnesium, and water	stripes, lack of right-angle cleavage and shiny appearance distinguish it from hornblende
hexagonal prisms, ometimes grooved		beryllium, aluminum silicate $Be_3Al_2(Si_6O_{18})$	clear blue-green beryl is the gem aquamarine; deep green is the emerald
rounded, elongated ;rains			
ıll prisms		aluminum oxide Al_2O_3	mica often coats the surface; when corundum is clear red it is ruby; all other colors are sapphire

Appendix C

IDENTIFYING CAPE-ISLAND ROCKS*

The following tables present the rocks according to their three major geologic divisions. These three divisions are based upon the geologic processes through which a rock may have passed.

Igneous Rocks (Table R1) originally were masses of molten lava or magma which, upon cooling, crystallized into mineral arrangements.

Sedimentary Rocks (Table R2) are detrital, organic or chemical substances which have settled out from suspension or solution to form layered deposits.

Metamorphic Rocks (Table R3) are any rocks (either igneous or sedimentary or metamorphic) which have undergone changes in composition and appearance by heat and/or pressure.

* Two excellent field guides which describe and picture common rocks are Pough's Peterson Field Guide and the Golden Nature Guide, both listed in the Bibliography, Part II.

Table R1

Igneous Rocks Most Likely to Be Found on the Cape and Islands

texture	light in color and light in weight		intermediate in color and weight	dark in color and heavy	very dark and heavy
	major minerals		*major minerals*	*major minerals*	*major minerals*
	quartz feldspar mica hornblende	feldspar mica hornblende (no quartz)	feldspar hornblende black mica	black mica pyroxene olivine feldspar (small amount)	olivine pyroxene
extremely large grains	PEGMATITE likely to contain beryl or other gems				
mosaics of easily visible interlocked grains	GRANITE	SYENITE	DIORITE	GABBRO	PERIDOTITE
tiny grains too small to distinguish	FELSITE light-colored; tan; gray; pink; green			BASALT dark-colored; black	

TABLE R2

Sedimentary Rocks Most Likely to Be Found on the Cape and Islands

Description	*Name*	*Where Found*
compact layers of dust-size grains; moldable when wet, hard when dry	CLAY	a. among Cretaceous sediments of Marthas Vineyard b. interstadial beds such as Gardiners Clay c. in tills here and there
thin consolidated layers of mud-size particles; brown; gray; green; black; red	SHALE	minor amounts among drift rocks and beach pebbles
layers of consolidated or nonconsolidated sand-size particles:		
1. particles primarily quartz; brown; red; gray; white; yellow	1. SAND or SANDSTONE	1a. beach and dune sands 1b. interstadial sands such as Nantucket's Sankaty Sand 1c. as fragments of consolidated sandstone among drift rocks and beach pebbles (not common)
2. particles include glauconite; green, may be red on surface	2. GREENSAND	2. Tertiary beds on Marthas Vineyard
gravel-size particles of quartz or other minerals (layers not always visible)		
1. particles unconsolidated; all colors	1. GRAVEL	1a. glacial outwash deposits 1b. beaches
2. particles consolidated; all colors	2. CONGLOMERATE	2. as fragments among drift rocks and beach pebbles
fragments of all sizes	TILL	frontal and ground moraines
gray, fine-grained; fizzes in vinegar or other acid when surface scratched; mainly calcite	LIMESTONE	as fragments among drift rocks and beach pebbles (not common)
brown spongy plant remains	1. LIGNITE 2. PEAT	1. in Cretaceous deposits, Marthas Vineyard 2. scattered through glacial deposits and currently forming in swamps

TABLE R3

Metamorphic Rocks Most Likely to Be Found on the Cape and Islands

			Name	Description	Occurrence
metamorphism changes:	clay; or shale	into	SLATE	very fine-grained; breaks into thin slabs with perfectly flat surfaces; gray; black; red; green	not common but occurs in minor amounts among drift rocks and beach pebbles
metamorphism changes:	sandstone	into	QUARTZITE	*very* hard, compact crystalline mass of quartz; white; gray	quite common among drift rocks and beach pebbles
metamorphism changes:	coarse-grained igneous rocks (as granite); or conglomerate; or gneiss	into	GNEISS	strongly banded with layers of different colors; minerals and grain size similar to those of coarse-grained igneous rocks; variously colored	quite common among drift rocks and beach pebbles
metamorphism changes:	fine-grained igneous rocks; or shale; or schist	into	SCHIST	splits easily along shiny, crinkly surfaces; often with mica, talc, chlorite, garnet or serpentine; variously colored	common among drift rocks and beach pebbles
metamorphism changes:	limestone	into	MARBLE	crystalline masses of calcite; fizzes when vinegar or other acid is applied; often streaked with graphite; white; gray; pink	not common but occurs among drift rocks and beach pebbles

Appendix D

AVAILABLE MAPS AND CHARTS

I. *U. S. Geological Survey Topographic Maps*

These are the most interesting and useful maps available for geologic rambling.

All of the following maps together constitute a very large, complete map of the Cape-Island region.

Each shows relief and cultural features. Some have green tint denoting woodlands. Each covers 7½′ of latitude and 7½′ of longitude, at a scale of: 1″=2000′.

NANTUCKET:

Tuckernuck Island
Nantucket
Siasconset
Great Point

MARTHAS VINEYARD:

Naushon Island
Squibnocket
Vineyard Haven
Tisbury Great Pond
Edgartown

CAPE COD:

Provincetown
North Truro
Wellfleet
Orleans
Chatham
Monomoy Point
Harwich
Dennis
Hyannis
Sandwich
Cotuit
Sagamore
Pocasset
Falmouth
Onset
Woods Hole

The above maps are available from: U. S. Geological Survey, Washington 25, D.C., and from some private map dealers.

II. *U. S. Coast and Geodetic Survey Charts*

These are designed for navigational purposes. They indicate depth of offshore waters, locations of shoals and rocks, harbors and jetties, and conspicuous features of the nearby landscapes.

Charts 1208, 1209, and 1210, when put together, form a large, complete map of Cape-Island waters.

Scales range from: 1″ = 200′, to 1″ = about 80 miles

1208 – waters around Cape Cod, exclusive of the Woods Hole region
1209 – waters around Nantucket Island and eastern Marthas Vineyard, including the Nantucket Shoals
1210 – waters around western Marthas Vineyard and Buzzards Bay
249 – Falmouth Harbor and Buzzards Bay
250 – Monomoy Shoals
257 – Chatham Roads
258 – Bass River and Hyannis Harbor
259 – Cotuit Bay region
343 – Nantucket Harbor
346 – waters around eastern Marthas Vineyard
347 – Vineyard Haven Harbor
348 – Woods Hole
580 – Provincetown Harbor

The above charts are available from: U. S. Coast and Geodetic Survey, Washington 25, D.C., and from some private map dealers.

Glossary

ALEWIFE. *Pomolobus pseudoharengus,* a fish closely related to herring.
ALGAE. Simple plants occurring as fresh- or saltwater seaweeds.
ALLUVIUM. Stream, river and flood deposits.
AMBER. Preserved resin, yellow or red, brittle, sometimes containing insects.
AQUINNAH CONGLOMERATE. The earliest Pleistocene deposit (preglacial) on Marthas Vineyard.
BACKWASH. The seaward movement of water perpendicular to the shore after waves have broken on a beach.
BAR. An elongate sand deposit built by currents parallel to a coast or at the mouth of a bay or harbor.
BAR, BAYHEAD. A sand deposit built by currents and waves and lying along the concave shores of a bay.
BARNSTABLE COUNTY. That county of Massachusetts which consists for the most part of Cape Cod.
BARRIER BEACH. An elongate sand deposit built by currents and waves parallel to a coast and generally forming a lagoon behind it.
BASALT. A dark, heavy, fine-grained igneous rock.
BLACK GRASS. *Juncus gerardi,* a beach grass which grows shoreward of the marsh region.
BOG IRON ORE. Low-grade iron ore formed by simple plants in swamps, especially in glaciated regions, from iron solutions originating in the nearby ground.
BREAKWATER. A wall, generally of rock, built parallel to shore, usually to provide calm anchorage for boats.
BUZZARDS BAY LOBE. That glacial lobe of the late Wisconsin glaciations which moved from southern Massachusetts and northern Rhode Island, building up the western Marthas Vineyard moraine and Cape Cod's Buzzards Bay moraine.
BUZZARDS BAY MORAINE. The Cape Cod recessional moraine which runs from Bourne to Woods Hole.
CAPE COD BAY LOBE. That glacial lobe of the late Wisconsin glaciations which moved southeast across Cape Cod Bay, building up the eastern Vineyard moraine, Nantucket's moraine, and Cape Cod's Sandwich moraine, and, together with the South Channel Lobe, building Cape Cod's interlobate moraine.
CHALCEDONY. Very fine-grained quartz with a waxy luster; includes agate, flint.
CHERT. Very fine-grained quartz, usually found as gray or black nodules.
CLAY. Very fine-grained sediment, usually plastic when wet.
CLAY POUNDS. That region of the ocean cliffs near Cape Cod Light at North Truro, Cape Cod, underlain by massive blue clay.

COASTAL PLAIN. The low seaboard plain of some parts of the continents, representing the sea floor during a former higher sea level.

CONGLOMERATE. A sediment consisting of coarse pebbles held together in some natural cementing matrix.

CONIFER. An evergreen (usually), cone-bearing tree.

CONTINENTAL SHELF. The flat, gently sloping edge of the continent, covered by shallow seawaters.

CONTINENTAL SLOPE. The submarine precipice between the continental shelf and the floor of the deep sea.

CRETACEOUS. That period in earth history, 70 to 135 million years ago, characterized by the development of flowering plants.

CRINOID. A sea animal related to starfish; it attaches itself to the sea floor by a segmented stem on which there is a cup-shaped body with radial arms.

CURRENT RIPPLES. Ripple marks in sand caused by moving currents, and leaning asymmetrically along the direction of current movement.

DELTA. A triangular-shaped alluvial deposit at the mouth of a river.

DEVONIAN. That period in earth history, 350 to 400 million years ago, characterized by early land plants.

DIATOM. A small one-celled or colonial water-dwelling plant with opaline hard parts.

DIKE. A long wall, usually of rock, built perpendicular to a coast to stop currents and thus protect a harbor.

DIORITE. An igneous rock midway between granite and basalt in appearance and composition.

DRIFT. Any sort of glacial deposit.

DUKES BOULDER BED. An early glacial deposit (Jameco substage) on Marthas Vineyard, no longer exposed.

DUKES COUNTY. That county of Massachusetts which consists of Marthas Vineyard, No Mans Land and the Elizabeth Islands.

EELGRASS. *Zostera marina,* a flowering perennial plant with green blades, which grows usually in the zone between the tides in relatively quiet waters.

EROSION. The wearing away of the land by water, wind, or ice.

ERRATIC. A glacially transported boulder, of a rock type usually differing from the rocks on which it is found.

ESTUARY. A partly fresh, partly salt river affected by the tides of the sea.

FLAGSTONE. A rock, usually a sandstone, which breaks into slabs.

FOSSIL. Any remains of a living thing or residue of a life process from the geologic past.

FOX GRASS. *Spartina patens,* a marsh grass which grows near the high-tide line.

GARDINERS CLAY. The interstadial marine clay on Marthas Vineyard, Cape Cod and Long Island, deposited between the Jameco and Manhasset glacial substages.

GASTROLITH. A stone from the stomach of an animal.

GEORGES BANK. The submarine hill reaching 150 miles northeast from the eastern end of the Nantucket Shoals.

GLACIAL LOBES. Finger-shaped masses of glacial ice caused by the splitting apart of an ice sheet around irregularities in the terrain over which it passes.

GLACIAL STAGE. A major period of glaciation, such as the Wisconsin, within a geologic epoch, such as the Pleistocene.

GLACIER. A mass of ice, formed of the unmelted snow of many seasons, which moves under its own weight.

GLACIO-ALLUVIAL DEPOSITS. Deposits of boulders, pebbles, sand, silt or clay, carried and dropped by meltwater streams from a glacier.

GLAUCONITE. A hydrated potassium mineral formed in shallow seas.

GLOBIGERINA. One-celled marine animals with minute calcareous shells.

GNEISS. A metamorphic rock, banded and coarse-textured.

GRANITE. A coarse-grained igneous rock with quartz and feldspar.

GRAPHIC GRANITE. A type of granite in which the quartz is so arranged within the feldspar as to resemble cuneiform writing.

GRAPTOLITE. An extinct colonial animal, the remains of which usually are carbon films shaped like thin, saw-toothed bands.

GREAT PLAINS. On Marthas Vineyard a local name for the outwash plain.

GREENSAND. A sandy deposit consisting mainly of the green mineral glauconite.

GROIN. A wall, usually of rock, built perpendicular to the shore to deflect currents and cause them to drop their sand. Usually groins are built to improve and enlarge beaches.

GROUNDWATER. Water found in pore spaces and cracks within the earth's upper crust.

GULF STREAM. The current which flows north in the Atlantic Ocean along the east coast of North America to south of Cape Cod, and thence eastward to Europe.

HARBOR HILL MORAINE. Long Island's recessional moraine. It runs along the island's north shore to Orient Point and continues eastward as Cape Cod's Buzzards Bay and Sandwich moraines.

HARDPAN. A subsoil deposit of clay or till, made hard and cohesive by cementing materials leached from above and carried downward by water.

HEMPSTEAD GRAVEL. One of the outwash gravels of the Manhasset substage on Marthas Vineyard.

HEROD GRAVEL. One of the outwash gravels of the Manhasset substage on Marthas Vineyard.

HIGH PLAINS OF WELLFLEET-TRURO. The central, highest and most irregular topographic division of Cape Cod's interlobate moraine.

HOLLOW. A low portion of the cliffs along the Lower Cape's coast, formed where an east-west valley meets the shore. The word is used as well for some of the valleys.

HORSESHOE CRAB. *Limulus polyphemus,* a mud-scavenging marine animal related to spiders.

IGNEOUS. Pertaining to rocks which originated as molten material from within the earth's crust.

INTERGLACIAL STAGE. A major interval within the Pleistocene, separating successive glacial stages.

INTERLOBATE MORAINE. The sandy drift deposits which make up most of the Lower Cape.

INTERSTADIAL INTERVAL. A minor interval within a glacial stage, characterized by short-term or local glacial retreat.

INTERTILL. A nonglacial sediment laid down between two successive tills during an interstadial interval.

JACOB SAND. The interstadial sand capping the Gardiners Clay on Marthas Vineyard.

JAMECO SUBSTAGE. The earliest known Wisconsin Stage glacial substage represented in the Cape-Island region.

JETTY. A wall, usually of rock, built perpendicular to a coast to deflect current-carried sand and protect a harbor.

KAME. A cone-shaped hill of layered glacial drift.

KAOLIN. A white clay mineral used for china, pottery, etc.

KETTLE HOLE. A depression in a glacial terrain, caused by the onetime presence of a block of ice broken from a retreating glacier and covered and surrounded by debris.

KNOB-AND-KETTLE. A type of glacial topography characterized by irregular hills and depressions.

LABRADOR CURRENT. That current in the Atlantic which flows from Labrador southward and swings east just north of Cape Cod.

LAGOON. A shallow body of water partly or wholly separated from the sea by a narrow sand barrier.

LAVA. Molten rock, from a volcanic vent, which spills onto the land surface and there solidifies.

LIGNITE. Very low-grade coal, often with plant matter still evident.

LONGSHORE CURRENT. A current, flowing parallel to a coast, set up by waves which approach shore at an angle.

LOWER CAPE. On Cape Cod, a local name for the north-south portion of the peninsula.

MAGMA. Molten rock which solidifies in chambers beneath the earth's surface.

MANETTO FORMATION. A bed of stony blue clay from the Jameco substage on Marthas Vineyard.

MANHASSET SUBSTAGE. The next-to-last glaciation within the Wisconsin Stage in southeastern New England.

MARBLE. A metamorphic rock consisting almost entirely of recrystallized limestone.

MICA. A mineral which splits into paper-thin flakes.

MOLLUSC. An invertebrate marine creature, usually shelled; the group includes clams and snails.

MONTAUK TILL. The till left on Marthas Vineyard during the Manhasset substage.

MOORS, THE. On Nantucket, the local name for the outwash plain.

MORAINE. A ridge of material deposited along the front or sides of, or beneath, a glacier.

MORAINE, FRONTAL. The glacial deposits laid down along the front margin of an ice sheet or lobe.

MORAINE, GROUND. The glacial deposits laid down beneath a moving ice sheet.

MORAINE, INTERLOBATE. A moraine built up along the edge between two adjoining glacial lobes.

MORAINE, RECESSIONAL. The glacial deposits laid down by an ice sheet as it retreats.

MORAINE, TERMINAL. The glacial deposits laid down at the farthest limit of a glacier's advance.

MOSHUP TILL. The earliest known till (Jameco substage) on Marthas Vineyard.

NORTH TRURO PLAINS. The northernmost topographic division of Cape Cod's interlobate moraine.

OPAL. Formless, hydrated silica.

OSCILLATION RIPPLES. Wave-produced, symmetrical, peaked ripple marks on a sandy, shallow sea floor.

OUTWASH. The deposits of sand and gravel carried from ice and sorted by meltwater streams.

OUTWASH PLAIN. A sand-gravel plain formed by the coalescence of individual outwash deposits.

PAMET. A stream bed covered with glacial drift.

PERIDOTITE. A coarse-grained rock rich in iron and magnesium.

PITTED PLAIN. An outwash plain such as that of Cape Cod, with abundant kettle holes.

PLAINS OF EASTHAM. The very flat part of the southern topographic division of Cape Cod's interlobate moraine.

PLEISTOCENE. That epoch in earth history, 600,000 to 10,000 years ago, characterized by widespread glaciation.

PLYMOUTH INTERLOBATE MORAINE. The recessional moraine which runs north along the Massachusetts mainland coast and through Plymouth.

PODZOL. An acid soil of cool, humid climates.

QUARTZ. A mineral form of silica, widely distributed in rocks.

RED GROUND. A local name for deposits of red Cretaceous clay around Edgartown, Marthas Vineyard.

RIP. The zone within a breaker line at which water of the surf converges to return seaward.

ROCK FLOUR. Very fine-grained rock fragments ground down by transportation in a glacier.

RONKONKOMA MORAINE. Long Island's terminal moraine. It runs along the central part of the island to Montauk Point, and continues eastward as the moraine on the Islands.

SALT THATCH. *Spartina stricta,* a marsh grass which grows between the eelgrass and fox grass zones.

SANDWICH MORAINE. The Cape Cod recessional moraine which runs eastward from Sagamore to Orleans.

SANKATY SAND. The interstadial deposit of fossiliferous sand on Nantucket.

SCENERY-MAKER, THE. The final glacier on the Cape and Islands; the sculptor of the landscape.

SCHIST. A metamorphic rock which can be split into crinkly plates.

SEDIMENTATION. The building up of deposits of detrital, organic, residual and precipitated materials.

SILICIFICATION. The process of introducing silica as a rock cement or an infiltration or replacement of organic tissues or of other minerals.

SOUTH CHANNEL LOBE. That glacial lobe of the late Wisconsin glaciations which moved south from Maine, east of the Lower Cape, building up the Nantucket Shoals and Georges Bank, and, together with the Cape Cod Bay Lobe, building Cape Cod's interlobate moraine.

SOUTH SEA. On Cape Cod a local name for the outwash plain.

SPICULE. A small, siliceous, needle-like body in a sponge, the part most likely to be found fossilized.

SPIKE-GRASS. *Distichlis spicata,* a marsh grass which grows in the fox grass zone.

SPIT. A deposit of current-built sand, attached to the land where the coast changes direction.

STADIA. Glacial substages.

STRATIFICATION. The arrangement of sedimentary rocks in beds or strata. In an undisturbed sequence the older are at the bottom.

STRIATION. A scratch on a rock resulting from glacial abrasion.

SUBGLACIAL DEPOSIT. A deposit such as ground moraine left beneath the body of a glacier as it moves.

SUBSTAGE. A subdivision of a glacial stage, characterized by local or short-term advances of the ice.

TABLELANDS OF NAUSET. That part of the Eastham Plains characterized by a very flat terrain.

TALUS. An accumulation at the foot of a cliff of debris which fell down as a result of weathering and erosion.

TERTIARY. That period in earth history, 70,000,000 to 600,000 years ago, characterized by the dominance of mammals.

TILL. An unlayered glacial deposit consisting of heterogeneous rock materials and boulders.

TILL 1. The upper part of the two tills of the final Wisconsin glaciation in Cape Cod's interlobate moraine.

TILL 1-a. The lower part of the two tills of the final Wisconsin glaciation in Cape Cod's interlobate moraine.

TILL 2: TILL 3: TILL 4. The three tills of the Manhasset substage of the Wisconsin glaciation in Cape Cod's interlobate moraine. Till 2 is the highest, Till 4 the lowest.

UP-ISLAND. On Marthas Vineyard, a local designation for the western part of the island.

UPPER CAPE. On Cape Cod, a local name for the east-west section of the peninsula.

VENTIFACTS. Stones shaped, pitted and frosted by the bombardment of wind-borne sand.

WATER TABLE. The level in the ground beneath which the ground is saturated with water held within its pore spaces and cracks.

WEATHERING. The alteration and decay of rocks by atmospheric agencies.

WEYQUOSQUE FORMATION. The outwash on Marthas Vineyard from the Jameco substage of the Wisconsin glaciation.

WISCONSIN STAGE. The final glacial stage within the Pleistocene Epoch. All of the Cape-Island ice deposits arrived in the Wisconsin.

Bibliography

I. A Comprehensive Bibliography of Cape-Island Geology

ALLEN, G. M. "The Whalebone Whales of New England." *Memoirs of the Boston Society of Natural History,* v. 8, No. 2, 1916

AMOS, W. H. "The Life of a Sand Dune." *Scientific American,* v. 201, No. 1, pp. 91–99, 1959

ANONYMOUS. "A Description of Dukes County." Massachusetts Historical Society *Collections,* Second Series, v. 3, pp. 38–92, 1815

ANONYMOUS. "Notes on Nantucket." Massachusetts Historical Society *Collections,* Second Series, v. 3, pp. 19–38, 1815

BASSLER, R. S. "Geology and Paleontology of the Georges Bank Canyon; Part III: Cretaceous Bryozoan from Georges Bank." Geological Society of America *Bulletin,* v. 47, pp. 411–412, 1936

BAYLIES, W. "Description of Gay Head." American Academy of Arts & Sciences *Memoirs,* No. 2, pp. 150–155, 1793

BEAUMONT, A. "Geography of New England Soils [abstract]." Geological Society of America *Bulletin,* v. 52, No. 12, pt. 2, p. 2010, 1941

BERRY, E. "The Age of the Cretaceous Flora of Southern New York and New England." *Journal of Geology,* v. 23, pp. 608–618, 1915

BLAKE, W. P. "Notes of the Occurrence of Siderite at Gay Head, Mass." *Transactions of the American Institute of Mining Engineers,* v. 4, pp. 112–113, 1876

BLUNT, E. *American Coast Pilot.* 11th ed., 1827; 13th ed., 1837

BOWMAN, I. "Northward Extension of the Atlantic Preglacial Deposits." *American Journal of Science,* Fourth Series, v. 22, pp. 313–325, 1906

BROWN, T. C. "A New Lower Tertiary Fauna for Chappaquiddick Island, M. V." *American Journal of Science,* Fourth Series, v. 20, pp. 229–238, 1905; *Science,* v. 21, pp. 990–991, 1905

BRYAN, K. "Wind-worn Stones or Ventifacts—a Discussion and Bibliography." National Resources Council, *Report of the Committee on Sedimentation,* pp. 29–50, 1929–1930

BRYAN, K. "New Criteria Applied to the Glacial Geology of Southeastern Mass. [abstract]." Geological Society of America *Bulletin,* v. 43, p. 176, 1932

BRYSON, J. "The Glacial Geology of Marthas Vineyard Compared with that of Long Island." *American Geologist,* v. 11, pp. 210–212, 1893

CHADWICK, G. H. "Glacial Molding of the Gulf of Maine [abstract]." *Earth Science Digest,* v. 4, No. 6, p. 15, 1950

CHASE, P.; WALCOTT, H.; SARGENT, C.; WIGGLESWORTH, G.; ELIOT, C. "Report of the Trustees of Public Reservations on the Subject of the Provincelands." Massachusetts House, *1st Annual Report,* No. 339, Appen. III, 1893

CHUTE, N. E. "Geology of the Coastline between Point Gammon and Monomoy

Point, Cape Cod, Mass." Commonwealth of Mass., Dept. of Public Works, Sp. Paper 1, pp. 1–26, 1939

CLARK, W. B. "The Geologic Features of Gay Head, Mass." John Hopkins University *Circular*, No. 10, p. 28, 1890

COMMONWEALTH OF MASSACHUSETTS. "Population and Resources of Cape Cod." Department of Labor and Industries, 1922

CROSBY, I. B.; LOUGEE, R. J. "Glacial Marginal Shores and the Marine Limit in Mass." Geological Society of America *Bulletin*, v. 45, pp. 441–462, 1934

CROSBY, W. O. "On the Occurrence of Fossiliferous Boulders in the Drift of Truro, on Cape Cod, Mass." Boston Society of Natural History *Proceedings*, v. 20, pp. 136–140, 1879

CROSBY, W. O.; LAFORGE, L. "Contributions to the Hydrology of the Eastern United States: Mass." U. S. Geological Survey, *Water Supply Paper 102*, pp. 94–118, 1904

CURRIER, W. L. "Tills in Eastern Mass. [abstract]." Geological Society of America *Bulletin*, v. 52, pp. 1895–1896, 1941

CURTIS, G. C.; WOODWORTH, J. B. "Nantucket, a Morainal Island." *Journal of Geology*, v. 7, pp. 226–236, 1899

CUSHMAN, J. A. "Miocene Barnacles from Gay Head, Mass., with Notes on *Balanus proteus* Conrad." *American Geologist*, v. 34, pp. 293–296, 1904

CUSHMAN, J. A. "Notes on the Pleistocene Fauna of Sankaty Head, Nantucket, Mass." *American Geologist*, v. 34, pp. 169–174, 1904

CUSHMAN, J. A. "Fossil Crabs of Gay Head Miocene." *American Naturalist*, v. 39, pp. 381–390, 1905

CUSHMAN, J. A. "Notes on Fossils Obtained at Sankaty Head, Nantucket, in July, 1905." *American Geologist*, v. 36, pp. 194–195, 1905

CUSHMAN, J. A. "The Pleistocene Deposits of Sankaty Head, Nantucket, and their Fossils." *Nantucket Maria Mitchell Assoc.*, Pub. 1, No. 1, pp. 1–21, 1906

CUSHMAN, J. A. "Geology and Paleontology of the Georges Bank Canyons; Part IV—Cretaceous and Late Tertiary Foraminifera." Geological Society of America *Bulletin*, v. 47, pp. 413–440, 1936

DALL, W. H. "Notes on the Miocene and Pliocene of Gay Head, Marthas Vineyard, Mass., and on the 'Land Phosphate' of the Ashley River District, S. C." *American Journal of Science*, Third Series, v. 48, pp. 296–301, 1894

DALL, W. H.; HARRIS, G. "The Neocene of North America." U. S. Geological Survey *Bulletin*, 84, pp. 32–38, 1892

DALY, R. A. "Oscillations of Level in the Belts Peripheral to the Pleistocene Ice-caps." Geological Society of America *Bulletin*, v. 31, pp. 303–318, 1920

DAVIS, W. "A Description of Sandwich, in the County of Barnstable." Mass. Historical Society *Collections*, First Series, v. 8, pp. 119–125, 1802

DAVIS, W. M. "Facetted Pebbles on Cape Cod." Boston Society of Natural History *Proceedings*, v. 26, pp. 166–175, 1893

DAVIS, W. M. "The Outline of Cape Cod." American Academy of Arts & Science *Proceedings*, v. 31, pp. 303–332, 1896

DEARBORN, H. A. "Sketch of the Mineralogy of Gay Head and of Bird Island." *Boston Journal of Philosophy & Arts*, v. 3, pp. 588–590, 1826

DESOR, E. "Drift Fossils from Nantucket, Mass." Boston Society of Natural History *Proceedings*, v. 3, pp. 79–80, 1848

DESOR, E. "Deposit of Drift Shells in the Cliffs of Sancati Island, or Nantucket." American Association for the Advancement of Science *Proceedings*, v. 1, pp. 100–101, 1849

DESOR, E.; CABOT, E. "On the Tertiary and More Recent Deposits in the Island of Nantucket." *Journal of the Geological Society of London Quarterly*, 5, pp. 340–344, 1849; Boston Society Natural History *Memoirs*, v. 6, p. 252, 1866

DRAKE, C. L.; WORZEL, J. L.; BECKMASS, W. C. "Geophysical Investigations in the Emerged and Submerged Atlantic Coastal Plain, Part IX, the Gulf of Maine." Geological Society of America *Bulletin*, v. 65, pp. 957–970, 1954

EMERSON, B. K. "Geology of Massachusetts and Rhode Island." U. S. Geological Survey *Bulletin* 597, 1917

EMILIANI, C. "Ancient Temperatures." *Scientific American*, vol. 198, No. 2, pp. 54–63, Feb., 1958

EWING, M.; CRARY, A. P.; RUTHERFORD, H. M. "Geophysical Investigations in the Emerged and Submerged Atlantic Coastal Plain, Part I, Methods and Results." Geological Society of America *Bulletin*, v. 48, pp. 753–802, 1937

EWING, M.; WORZEL, J. L.; STEENLAND, N. C.; PRESS, F. "Geophysical Investigations in the Emerged and Submerged Atlantic Coastal Plain, Part V, Woods Hole, New York and Cape May Sections." Geological Society of America *Bulletin*, v. 61, pp. 877–897, 1950

FAIRCHILD, H. L. "Postglacial Uplift of Southern New England." Geological Society of America *Bulletin*, v. 30, pp. 597–636, 1919

FINCH, J. "Geological Essay on the Tertiary Formations in America." *American Journal of Science and Arts*, v. 7, pp. 31–43, 1823

FLINT, R. F. "How Many Glacial Stages are Recorded in New England?" *Journal of Geology*, v. 43, No. 7, pp. 771–777, 1935

FLINT, R. F. "Probable Wisconsin Substages and Late Wisconsin Events in Northeastern United States and Southeastern Canada." Geological Society of America *Bulletin*, v. 64, pp. 897–919, 1953.

FLINT, R. F. *Glacial and Pleistocene Geology*. New York: John Wiley & Sons, 1957.

FOLGER, W. "A Topographical Description of Nantucket." Massachusetts Historical Society *Collections*, First Series, v. 3 (for year 1794), pp. 153–155, 1810

FREEMAN, J. "A Description of Chatham, in the County of Barnstable." Massachusetts Historical Society *Collections*, v. 8, pp. 142–154, 1802

FREEMAN, J. "A Description of Dennis, in the County of Barnstable." Massachusetts Historical Society *Collections*, v. 8, pp. 129–140, 1802

FREEMAN, J. "A Description of the Eastern Coast of the County of Barnstable from Cape Cod, or Race Point, in Lat. 42° 5′, to Cape Mallebarre, or the Sandy Point of Chatham, in Lat. 41° 33′, Pointing Out the Spots, on which the Trustees of the Humane Society Have Erected Huts, and Other Places Where Shipwrecked Seamen May Look for Shelter." Massachusetts Historical Society *Collections*, v. 8, pp. 110–119, 1802

FREEMAN, J. "Description and History of Eastham, in the County of Barnstable." Massachusetts Historical Society *Collections*, v. 8, pp. 154–186, 1802

FREEMAN, J. "Note on Falmouth, in the County of Barnstable." Massachusetts Historical Society *Collections*, First Series, v. 8, pp. 127–129, 1802

FREEMAN, J. "A Description of Provincetown, in the County of Barnstable." Massachusetts Historical Society *Collections*, First Series, v. 8, pp. 196–202, 1802

FREEMAN, J. "A Description of Orleans, in the County of Barnstable." Massachusetts Historical Society *Collections*, First Series, v. 8, pp. 186–195, 1802

FREEMAN, J. "Note on the Southern Precinct of Harwich, in the County of Barnstable." Massachusetts Historical Society *Collections*, First Series, v. 8, pp. 141–142, 1802

FREEMAN, J. "Topographical Description of Truro." Massachusetts Historical Society Collections, First Series, v. 3, (for year 1794), pp. 195–201, 1810

FREEMAN, J. "A Description of Mashpee, in the County of Barnstable." Massachusetts Historical Society *Collections*, Second Series, v. 3, pp. 1–17, 1815

FULLER, M. L. "Clays of Cape Cod, Mass." U. S. Geological Survey *Bulletin*, 285, pp. 432–441, 1905

FULLER, M. L. "Glacial Stages in Southeastern New England and Vicinity." *Science*, New Series, v. 24, pp. 467–469, 1906

FULLER, M. L. "The Geology of Long Island." U. S. Geological Survey *Professional Paper 82*, 1914

GALE, B. "Report on the Geologic Features on a Portion of Cape Cod, Mass." *National Park Service Field Investigation Report*, 1958.

GOLDSMITH, E. "On Amber Containing Fossil Insects." Philadelphia Academy of Natural Science *Proceedings*, pp. 207–208, 1879

GRABAU, A. W. "The Sand and Gravel Plains of Truro, Wellfleet and Eastham, Cape Cod." *Science*, New Series, v. 5, pp. 334–335, 1897

GULLIVER, F. P. "Nantucket Shorelines II." Geological Society of America *Bulletin*, v. 15, pp. 507–522, 1904

GULLIVER, F. P. "Wauwinet-Coskata Tombolo, Nantucket, Mass. [abstract]." British Association for the Advancement of Science, *Report* 79, p. 536

GULLIVER, F. P. "Nantucket Shorelines IV [abstract]." Geological Society of America *Bulletin*, v. 20, p. 670, 1909

HASKINS, H.; KNOTT, S. "Geophysical Investigation of Cape Cod Bay, Mass., Using the Continuous Seismic Profile." *Journal of Geology*, v. 69, No. 3, pp. 330–340, 1961

HITCHCOCK, E. "Notices of the Geology of Marthas Vineyard and the Elizabeth Islands." *American Journal of Science and Arts*, v. 7, pp. 240–248, 1824

HITCHCOCK, E. "Report on the Geology of Massachusetts Examined Under the Direction of the Government of that State During the Years 1830 and 1831." *American Journal of Science*, v. 22, pp. 1–70, 1832

HITCHCOCK, E. *Report on the Geology, Mineralogy, and Zoology of Massachusetts*. Amherst, Mass., 1833

HITCHCOCK, E. *Report on the Geology, Mineralogy, Botany and Zoology of Massachusetts*." Amherst, Mass., 1835

HITCHCOCK, E. "Report on a Reexamination of the Economical Geology of Massachusetts." *House Document* 52, Boston, 1838

HITCHCOCK, E. *Final Report on the Geology of Massachusetts.* In 2 volumes, Northampton, 1841

HOLLICK, A. "Observations on the Geology and Botany of Marthas Vineyard." New York Academy of Sciences *Transactions,* 13, pp. 8–22, 1893

HOLLICK, A. "Dislocations in Certain Portions of the Atlantic Coastal Plain and their Probable Cause." New York Academy of Sciences *Transactions,* 14, pp. 8–20, 1895

HOLLICK, A. "Geological Notes, Long Island and Nantucket." New York Academy of Sciences *Transactions,* 15, pp. 3–10, 1896

HOLLICK, A. "Marthas Vineyard Cretaceous Plants [abstract]." Geological Society of America *Bulletin,* v. 7, pp. 12–14, 1896; *American Geologist,* v. 16, p. 239; *Science,* New Series, v. 2, p. 281

HOLLICK, A. "Geological and Botanical Notes, Cape Cod and Chappaquiddick Island, Mass." New York Botanical Garden *Bulletin,* v. 2, pp. 381–407, 1902

HOLLICK, A. "The Cretaceous Flora of Southern New York and New England." U. S. Geological Survey Monograph 50, 1906

HORNER, N. "Late Glacial Marine Limit in Massachusetts." *American Journal of Science,* v. 17, pp. 123–145, 1929

HOUGH, J. L. "Suggestion Regarding the Origin of Rock Bottom Areas in Massachusetts Bay." *Journal of Sedimentary Petrology,* v. 2, p. 131, 1932

HOUGH, J. L. "Sediments of Buzzards Bay, Mass." *Journal of Sedimentary Petrology,* v. 10, No. 1, pp. 19–32, 1940

HOUGH, J. L. "Sediments of Cape Cod Bay, Mass." *Journal of Sedimentary Petrology,* v. 12, No. 1, pp. 10–30, 1942

HOWE, O. "The Hingham Red Felsite Boulder Train." *Science,* v. 84, pp. 394–396, 1936

HYYPPÄ, E. "On the Pleistocene Geology of Southern New England." Societa Geographica Fenniae; *Acta Geographica,* v. 14, pp. 155–225, 1955

JAHNS, R. H. "Sheet Structure in Granites; its Origin and Use as a Measure of Glacial Erosion in New England." *Journal of Geology,* v. 51, pp. 71–98, 1943

JEFFREY, E. C. "A New Prepinus from Marthas Vineyard." Boston Society of Natural History *Proceedings,* v. 34, No. 10, pp. 333–338, 1910

JOHNSON, D. *The New England-Acadian Shoreline.* New York: John Wiley and Sons, 1925

JONES, W.; LUCKE, J. "Nantucket Shorelines [abstract]." Geological Society of America *Bulletin,* v. 62, No. 12, pt. 2, p. 1453, 1951

JULIEN, A. "On the Pebbles at Harwich (Cape Cod), Mass., and on Rude Arrowheads Found Among Them." *Science,* New Series, v. 26, pp. 831–832, 1907

KATZ, S.; EDWARDS, R. S.; PRESS, F. "Crustal Structure Beneath the Gulf of Maine." Columbia University Lamont Geological Observatory Technical Report on Seismology, No. 9, 1950

KNOX, A. "Micropaleontology and Geology of the Gay Head Cliffs, Mass." Geological Society of America *Bulletin,* v. 63, No. 12, pt. 2, p. 1271, 1952

KOONS, B. F. "Upon the Kettle Holes near Woods Hole, Mass." *American Journal of Science,* Third Series, v. 27, pp. 260, 264, 1884

KOONS, B. F. "Additional Notes on the Kettle Holes of the Woods Hole Region, Mass." *American Journal of Science,* Third Series, v. 29, pp. 480–486, 1885

LEET, L. "The Provincetown, Mass., Earthquake of April 23, 1932, and Data for Investigating New England's Seismicity." National Academy of Sciences *Proceedings,* v. 21, No. 6, pp. 308–313, 1935

LOUGEE, R. J. "Early Marine Stage of the Last Glaciation in Southern New England [abstract]." Geological Society of America *Bulletin,* v. 50, p. 1919, 1939

LUTZ, H. J. "The Nature and Origin of Layers of Fine-Textured Material in Sand Dunes." *Journal of Sedimentary Petrology,* v. 11, No. 3, pp. 105–123, 1941

LYELL, C. "On the Tertiary Strata of the Island of Marthas Vineyard in Mass." *Proceedings* of the Geological Society of London, v. 4, pp. 31–33, 1841–1843; *American Journal of Science,* v. 46, pp. 318–320, 1844

LYELL, C. "*Travels in North America.*" New York: pp. 203–207, 1845

MACY, Z. "Description of Nantucket." Massachusetts Historical Society *Collections,* 1st Ser., v. 3 (for year 1794), p. 158, 1810

MARINDIN, H. L. "Encroachment of the Sea Upon the Coast of Cape Cod, Mass., as Shown by Comparative Studies; Cross-sections of the Shores of Cape Cod Between Chatham and Highland Lighthouse." U. S. Coast Survey *Report,* pp. 403–407, 1889

MARINDIN, H. L. "On the Changes in the Shorelines and Anchorage Areas of Cape Cod (or Provincetown) Harbor as Shown by a Comparison of Surveys Made Between 1835, 1867, and 1890." U. S. Coast and Geodetic Survey, Appendix no. 8 to *Report* for 1891, pt. 2, pp. 283–288, 1891

MARINDIN, H. L. "On the Changes in the Ocean Shorelines of Nantucket Island, Mass., from a Comparison of Surveys Made in the Years 1846 to 1887 and in 1891." U. S. Coast and Geodetic Survey Report for 1892, pt. 2, pp. 243–252, 1894

MATHER, K. "Glacial Geology in the Buzzards Bay Region and Western Cape Cod." Field Trip No. 4, *Guidebook for Field Trips in New England,* Geological Society of America, 1952

MATHER, K.; GOLDTHWAIT, R.; THIESMEYER, L. "Preliminary Report on the Geology of Western Cape Cod, Mass." Massachusetts Department of Public Works *Bulletin,* 2, No. 3, 1940

MATHER, K.; GOLDTHWAIT, R.; THIESMEYER, L. "Retreatal Stages of Wisconsin Ice in Southeastern Mass. [abstract]." Geological Society of America *Bulletin,* v. 52, No. 12, pt. 2, p. 2018, 1941

MATHER, K.; GOLDTHWAIT, R.; THIESMEYER, L. "Pleistocene Geology of Western Cape Cod, Mass." Geological Society of America *Bulletin,* v. 53, pp. 1127–1174, 1942

MELLEN, J. "Topographical Description of Barnstable." Massachusetts Historical Society *Collections,* 1st ser., v. 3, pp. 12–17, 1794

MERRILL, F. "Post Pliocene Deposits of Sankaty Head." New York Academy of Sciences *Transactions*, v. 15, pp. 10–16, 1896

MILLER, G. "The Beach Mouse of Muskeget Island." Boston Society of Natural History *Proceedings*, v. 27, pp. 75–87, 1897

MILNE, L.; MILNE, M. "The Eelgrass Catastrophe." *Scientific American*, v. 184, No. 1, pp. 52–56, 1951

MITCHELL, H. "Report Concerning Nauset Beach and the Peninsula of Monomoy." U. S. Coast Survey *Report* for 1871, Appendix 9, pp. 134–143, 1871

MITCHELL, H. "Report on Physical Surveys Made at Marthas Vineyard and Nantucket During the Summer of 1871." U. S. Coast Survey *Report* for 1869, pp. 254–259, 1872

MITCHELL, H. "Additional Report on the Changes in the Neighborhood of Chatham and Monomoy." U. S. Coast Survey *Report* for 1873, Appendix 9, pp. 103–107, 1873

MITCHELL, H. "Monomoy and Its Shoals." U. S. Coast and Geodetic Survey *Report*, pp. 22–23, 1886; Harbor and Land Commission of Massachusetts Annual *Report*, 1887

MURRAY, H. "Topography of the Gulf of Maine." Geological Society of America *Bulletin*, v. 58, pp. 153–196, 1947

NATIONAL PARK SERVICE, DEPARTMENT OF THE INTERIOR. *A Report on a Seashore Recreation Area Survey of the Atlantic and Gulf Coasts*. 1955

PACKARD, A. S. "A New Fossil Crab from the Miocene Greensand Bed of Gay Head, Marthas Vineyard, with Remarks on the Phylogeny of the Genus Cancer." American Academy of Arts and Sciences *Proceedings*, v. 26, pp. 1–9, 1900

PENROSE, R. "Nature and Origin of Deposits of Phosphate of Lime." U. S. Geological Survey *Bulletin*, 46, p. 78, 1888

RANDALL, W.; VINAL, W. "Report of a Biological Investigation on a Portion of Cape Cod, Mass." National Park Service, "A Field Investigation Report on a Proposed National Seashore, Cape Cod, Barnstable County, Mass.," 1958

SANFORD, S. N. "Fossils of Colorful Gay Head." Boston Society of Natural History *Bulletin*, 7, pp. 3–5, 1934

SAYLES, R. W. "Upper Till, Two Boulder Clays, and Interglacial Flora on Cape Cod, Mass." Geological Society of America *Bulletin*, v. 50, pp. 1931–1932, 1939

SAYLES, R. W.; KNOX, A. "Fossiliferous Tills and Intertill Beds of Cape Cod, Mass." Geological Society of America *Bulletin*, v. 54, pp. 1569–1612, 1943

SCHALK, M. "A Textural Study of the Outer Beach of Cape Cod." *Journal of Sedimentary Petrology*, v. 8, pp. 41–56, 1938

SCUDDER, S. "Post-Pliocene Fossils from the Bluff at Sankaty Head, Nantucket." Boston Society of Natural History *Proceedings*, v. 18, pp. 182–185, 1876

SCUDDER, S. "The Pine Moth of Nantucket." Massachusetts Society for the Promotion of Agriculture, Boston, 1883

SHALER, N. S. "Preliminary Report on Seacoast Swamps of the Eastern United States." U. S. Geological Survey Annual *Report*, 6, pp. 353–398, 1886

SHALER, N. S. "Report on the Geology of Marthas Vineyard." U. S. Geological Survey Annual *Report*, 7, pp. 297–363, 1888

SHALER, N. S. "Geology of Nantucket." U. S. Geological Survey *Bulletin*, 53, pp. 601–652, 1889

SHALER, N. S. "On the Occurrence of Fossils of Cretaceous Age on the Island of Marthas Vineyard, Mass." Harvard Museum of Comparative Zoology *Bulletin*, v. 16, pp. 89–97, 1889

SHALER, N. S. "Tertiary and Cretaceous Deposits of Eastern Mass." Geological Society of America *Bulletin*, v. 1, pp. 443–452, 1890

SHALER, N. S. "The Conditions of Erosion Beneath Deep Glaciers, Based Upon a Study of the Boulder Train from Iron Hill, Cumberland, R. I." Harvard Museum of Comparative Zoology *Bulletin*, v. 16, pp. 185–225, 1893

SHALER, N. S. "The Geologic History of Harbors." U. S. Geological Survey Annual *Report*, 13, pt. 2, pp. 93–209, 1893

SHALER, N. S. "Geology of the Cape Cod District." U. S. Geological Survey Annual *Report*, 18, pt. 2, pp. 497–593, 1897

SHALER, N. S.; WOODWORTH, J. B.; MARBUTT, C. F. "The Glacial Brick Clays of Rhode Island and Southeastern Massachusetts." U. S. Geological Survey Annual *Report*, 17, pt. 1, pp. 957–1004, 1896

SHALOWITZ, A. "Our Changing Coastline." *Journal of Geology*, v. 39, No. 1, pp. 1–10, 1940

SHEPARD, F. "Marine Beaches of the United States [abstract]." Geological Society of America *Bulletin*, v. 62, No. 12, pt. 2, pp. 1477–1478, 1951

SHEPARD, F.; TREFETHEN, J.; COHEE, G. "Origin of Georges Bank." Geological Society of America *Bulletin*, v. 45, pp. 281–302, 1934

SIMPKINS, J. "Topographical Description of Brewster." Massachusetts Historical Society *Collections*, First Series, v. 10, pp. 72–79, 1806

SMITH, SARA. *Nantucket; A Brief Sketch of its Physiography and Botany.* Knickerbocker Press, 1901

STEPHENSON, L. W. "Geology and Paleontology of the Georges Bank Canyons, pt. 2, Upper Cretaceous Fossils from Georges Bank." Geological Society of America *Bulletin*, v. 47, pp. 367–410, 1936

STETSON, H. "Geology and Paleontology of the Georges Bank Canyons, pt. 1, Geology." Geological Society of America *Bulletin*, v. 47, pp. 339–366, 1936

STETSON, H. "The Sediments of the Continental Shelf off the Eastern Coast of the United States." Massachusetts Institute of Technology and Woods Hole Oceanographic Institute, *Papers in Physical Oceanography and Meteorology*, v. 5, No. 4, 1938

STETSON, H. "The Sediments and Stratigraphy of the Eastern Coast Continental Margin, Georges Bank to Norfolk Canyon." Massachusetts Institute of Technology and Woods Hole Oceanographic Institute, *Papers in Physical Oceanography and Meteorology*, v. 11, No. 2, 1949

STEVENS, C.; CROSS, C.; PIPER, W. "The Cranberry Industry in Massachusetts." Massachusetts Department of Agriculture *Bulletin*, 157, 1957

STIMPSON, W. "On the Fossil Crab of Gay Head." *Boston Journal of Natural History*, pp. 583–589, 1863

TOWNSEND, C. "Notes on an Early Chart of Long Island Sound and its Approaches." U. S. Coast and Geodetic Survey *Report*, pp. 775–777, 1890

UHLER, P. "A Study of Gay Head, Marthas Vineyard." Maryland Academy of Science *Transactions,* New Series, v. 1, pp. 204–212, 1892

U. S. ARMY CORPS OF ENGINEERS. "Water Resources Development in Massachusetts," 1959

U. S. DEPARTMENT OF COMMERCE. *United States Coast Pilot; Atlantic Coast, Section B; Cape Cod to Sandy Hook,* U. S. Coast and Geodetic Survey, 5th ed., 1950

UPHAM, W. "Glacial History of the New England Islands, Cape Cod and Long Island." *American Geologist,* v. 24, pp. 72–92, 1899

UPHAM, W. "The Fishing Banks Between Cape Cod and Newfoundland." *American Journal of Science,* Third Series, v. 47, pp. 123–129, 1894; Boston Society of Natural History *Proceedings,* 26, pp. 42–48, 1893

UPHAM, W. "The Formation of Cape Cod." *American Naturalist,* v. 13, pp. 489–502, 552–565, 1879

VERRILL, A. "On the Post-Pliocene Fossils of Sankaty Head, Nantucket Island." Brief Contribution to Zoology from the Museum of Yale College, No. 36, *American Journal of Science,* Third Series, v. 10, pp. 364–375, 1875

VERRILL, A. "Occurrence of Fossiliferous Tertiary Rocks on the Grand Bank and Georges Bank." *American Journal of Science,* Third Series, v. 16, pp. 323–324, 1878

WARD, L. F. "Age of the Island Series." *Science,* New Series, v. 4, pp. 757–760, 1896

WEST, S. "A Letter Concerning Gay Head." American Academy of Arts and Sciences *Memoirs,* 2, pp. 147, 150; 1793

WESTGATE, J. "Reclamation of Cape Cod Sand Dunes." U. S. Department of Agriculture, Bureau of Plant Industries *Bulletin,* 65, 1904

WHITE, C. "The Cretaceous." U. S. Geological Survey Correlation Papers, *Bulletin,* 82, pp. 93, 94, 1891

WHITE, D. "On the Cretaceous Plants from Marthas Vineyard [abstract]." Geological Society of America *Bulletin,* v. 1, pp. 554–556, 1890; *American Journal of Science,* Third Series, v. 39, pp. 93–101, 1890

WHITING, H. "Report on the Special Study of Provincetown Harbor, Mass." U. S. Coast and Geodetic Survey *Report* for 1867, Appendix 12, pp. 149–157, 1867

WHITING, H. "Report on Shoreline Changes at Edgartown Harbor, Mass." U. S. Coast and Geodetic Survey *Report* for 1871, Appendix 17, pp. 262–265, 1871

WHITING, H. "Report of Changes in the Shoreline and Beaches of Marthas Vineyard as Derived from Comparisons of Recent with Former Surveys." U. S. Coast and Geodetic Survey *Report* for 1886, Appendix 9, pp. 263–266, 1887

WHITING, H. "Recent Changes in the South Inlet into Edgartown Harbor, Marthas Vineyard." U. S. Coast and Geodetic Survey *Report* for 1889, Appendix 14, pp. 459–460, 1890

WHITING, H.; MITCHELL, H. "Report Concerning Marthas Vineyard and Nantucket." U. S. Coast and Geodetic Survey *Report* for 1869, Appendix 15, pp. 236–259, 1869

WHITMAN, L. "A Topographical Description of Wellfleet, in the County of

Barnstable." Massachusetts Historical Society *Collections*, First Series, v. 3 (for year 1794), pp. 117–126, 1810

WHITMAN, L. "Account of the Creeks and Islands in Wellfleet and Observations on the Importance of Cape Cod Harbor." Massachusetts Historical Society *Collections*, First Series, v. 4, pp. 41–43, 1795

WILDER, B. "Evidence as to the Former Existence of Large Trees on Nantucket." American Association for the Advancement of Science *Proceedings*, p. 294, 1894

WILSON, J. H. "The Pleistocene Formations of Sankaty Head, Nantucket." *Journal of Geology*, v. 13, pp. 713–734, 1905

WILSON, J. H. *The Glacial History of Nantucket and Cape Cod.* New York: Columbia University Press, 1906

WOODWORTH, J. B. "Note on the Occurrence of Erratic Cambrian Fossils in the Neocene Gravels of the Island of Marthas Vineyard." *American Geologist*, v. 9, pp. 243–247, 1892

WOODWORTH, J. B. "Postglacial Aeolian Action in Southern New England." *American Journal of Science*, Third Series, v. 47, pp. 63–71, 1894

WOODWORTH, J. B. "Unconformities in Marthas Vineyard and Block Island." Geological Society of America *Bulletin*, v. 11, 1897

WOODWORTH, J. B. "Some Glacial Wash-plains of Southern New England." *Bulletin* of the Essex Institute, v. 29, pp. 71–119, 1898

WOODWORTH, J. B. "Glacial Origin of Older Pleistocene in the Gay Head Cliffs, with a Note on the Fossil Horse of that Section." Geological Society of America *Bulletin*, v. 8, pp. 459–460, 1900

WOODWORTH, J. B.; WIGGLESWORTH, E. "The Geology of Cape Cod and the New England Islands." Harvard Museum of Comparative Zoology *Memoirs*, 52, 1934

ZEIGLER, J. "Beach Studies in the Cape Cod Area Conducted During the Period, August 1953–April 1960." Woods Hole Oceanographic Institute (Technical Report), Reference No. 60–20, 1960

ZEIGLER, J.; TUTTLE, S. "Beach Changes Based on Daily Measurements of Four Cape Cod Beaches." *Journal of Geology*, v. 69, No. 5, 1961

ZEIGLER, J.; TUTTLE, S.; HAYES, C. "Beach Changes During Storms, Outer Cape Cod, Mass." *Journal of Geology*, v. 67, No. 3, pp. 318–336, 1959

ZEIGLER, J.; HOFFMEISTER, W.; GIESE, G.; TASHA, H. "Discovery of Eocene Sediments in Subsurface of Cape Cod." *Science*, v. 132, No. 3437, pp. 1397–1398, 1960

II. Related Readings of General Interest

ARCHER, GABRIEL. "The Relation of Captain Gosnold's Voyage to the North Part of Virginia, 1602." Massachusetts Historical Society *Collections for 1602*

BESTON, HENRY. *The Outermost House*, New York: Rinehart and Co., Inc., 1928, 1949

BRERETON, M. JOHN. "A Brief and True Relation of the Discovery of the North Part of Virginia." Massachusetts Historical Society *Collections for 1602*

CARSON, RACHEL. *The Sea Around Us*. New York: Oxford University Press, 1951

CHAMPLAIN, SAMUEL. *Voyages*. Book I, 1612

CREVECOEUR, J. HECTOR ST. JOHN DE. *Letters From an American Farmer*. 1782

DWIGHT, TIMOTHY. *Travels in New England and New York*. V. III, 1822

Flateyjarbok. An account of the old Viking Saga of Eric the Red. (The title is roughly translated as "Flat Island Book.")

HAY, JOHN. *The Great Beach*. New York: Doubleday and Co., Inc., 1963

HAY, JOHN. *The Run*. New York: Doubleday and Co., Inc., 1959

KITTREDGE, HENRY. *Cape Cod: Its People and Their History*. Boston: Houghton Mifflin Co., 1930

LINCOLN, JOSEPH C. *Cape Cod Yesterdays*. Boston: Little, Brown and Company, 1935

MOURT, C. "A Relation or Journal of a Plantation Settled at Plymouth in New England, and Proceedings Thereof." Massachusetts Historical Society *Collections*, v. 8, 1622, 1625

POHLE, FREDERICK. *The Vikings on Cape Cod*. Pictou, Nova Scotia: Pictou Advocate Press, 1957

POUGH, FREDERICK. *A Field Guide to Rocks and Minerals*. Peterson Field Guide Series, Boston: Houghton Mifflin Co., 1960

SHAY, EDITH; SHAY, FRANK. *Sand in their Shoes: A Cape Cod Reader*. Boston: Houghton Mifflin Co., 1951

SHEPARD, FRANCIS. *The Earth Beneath the Sea*. Baltimore: Johns Hopkins Press, 1959

SMITH, JOHN. "A Description of New England." Massachusetts Historical Society *Collections*, v. 6, Series 3, 1616

THOREAU, HENRY. *Cape Cod*. 1865

ZIM, HERBERT; INGLE, LESTER. *Seashores*, A Golden Nature Guide. New York: Golden Press, 1955

ZIM, HERBERT; SHAEFFER, PAUL. *Rocks and Minerals*, A Golden Nature Guide. New York: Golden Press, 1957

Index

INDEX

Barbara Blau Chamberlain's interest in the geology of Cape Cod and the Islands began when, in the summers during her undergraduate and graduate study, she visited her parents who are year-round residents at Brewster on the Cape. The local rocks there had been dropped by glaciers and, using microscopic and mineralogical analyses, she set about tracing their places of origin on the mainland and the sea floor. From these early studies developed an over-all interest in the geology of the area.

A graduate of Barnard College in New York City, she attended the University of Colorado in Boulder, and Columbia University, from which she received her Master's degree in geology. She has taught at Hunter College in New York City and at Vassar College.

A member of the Geological Society of America and the American Mineralogical Society, she presently lives in Salisbury, Vermont, where her husband teaches mathematics at the University of Vermont.